Biochemical Applications of Raman and Resonance Raman Spectroscopies

MOLECULAR BIOLOGY

An International Series of Monographs and Textbooks

Editors: BERNARD HORECKER, NATHAN O. KAPLAN, JULIUS MARMUR, AND HAROLD A. SCHERAGA

A complete list of titles in this series appears at the end of this volume.

Biochemical Applications of Raman and Resonance Raman Spectroscopies

P. R. Carey

Division of Biological Sciences
National Research Council of Canada
Ottawa, Canada

1982

ACADEMIC PRESS

A Subsidiary of Harcourt Brace Jovanovich, Publishers

New York   London
Paris   San Diego   San Francisco   São Paulo   Sydney   Tokyo   Toronto

7272·5916

CHEMISTRY

ACADEMIC PRESS, INC.
111 Fifth Avenue, New York, New York 10003

United Kingdom Edition published by
ACADEMIC PRESS, INC. (LONDON) LTD.
24/28 Oval Road, London NW1 7DX

Library of Congress Cataloging in Publication Data

Carey, P. R.
    Biochemical applications of Raman and resonance Raman
spectroscopies.

    (Molecular biology series)
    Bibliography: p.
    Includes index.
    1. Raman spectroscopy. 2. Raman effect, Resonance.
3. Biological chemistry--Technique. I. Title.
II. Series.
QP519.9.R36C37   574.19'285   82-6667
ISBN 0-12-159650-8   AACR2

PRINTED IN THE UNITED STATES OF AMERICA

82 83 84 85   9 8 7 6 5 4 3 2 1

Contents

Preface   ix

CHAPTER 1. **Introduction**

   I.   Raman Scattering   1
  II.   The Appearance of a Raman Spectrum   3
 III.   Basic Units   4
 IV.   Raman, Resonance Raman, and Infrared Spectroscopies   6
  V.   Advantages and Disadvantages of Raman Spectroscopy
       for Biochemical Studies   8

CHAPTER 2. **Principles of Raman Spectroscopy**

    I.   Introduction   11
   II.   The Vibrations of a Diatomic Molecule from a Classical
       Standpoint   11
  III.   A Classical Model for Raman Scattering   17
  IV.   Comparison with Infrared Spectroscopy   19
   V.   A Quantum Mechanical Picture of Raman Scattering   21
  VI.   The Vibrations of a Polyatomic Molecule   26
 VII.   Polarization Properties of Raman Scattering   39
VIII.   Raman Intensities   42

CHAPTER 3. **Experimental Raman Spectroscopy**

  I. Basic Optics of the Raman Experiment   48
 II. Sampling Techniques and Sample Problems   57
III. Sophisticated Techniques   63

CHAPTER 4. **Protein Conformation from Raman
and Resonance Raman Spectra**

  I. Protein Structure   71
 II. Introduction to Raman and Resonance Raman Studies
     of Proteins   76
III. Raman Studies of Proteins   76
 IV. Resonance Raman Studies of the Chromophores
     of the Polypeptide Chain Using Excitation below 300 nm   96

CHAPTER 5. **Resonance Raman Studies of Natural,
Protein-Bound Chromophores**

   I. Introduction   99
  II. Heme Proteins   100
 III. Vitamin B_{12} (Cobalamin)   119
  IV. Chlorophylls   122
   V. Carotenoids   125
  VI. The Visual Pigments and Bacteriorhodopsin   133
 VII. Flavin Nucleotides   140
VIII. Metalloproteins   143

CHAPTER 6. **Resonance Raman Labels**

  I. Introduction   154
 II. Drug–Protein Complexes and Other Enzyme-Inhibitor
     Systems   155
III. Antibody–Hapten Complexes   163
 IV. Permanently Labeled Sites—Arsanilazocarboxypeptidase A   166
  V. Enzyme–Substrate Reactions   169
 VI. Nucleic Acids and Membranes   183

CHAPTER 7. **Nucleic Acids and Nucleic Acid–Protein Complexes**

  I. The Structure of Nucleic Acids   184
 II. Polynucleotide Structure from the Normal Raman
     Spectrum   187
III. Applications to RNA and DNA Structures   195
 IV. Viruses and Other Nucleic Acid–Protein Complexes   198
  V. Resonance Raman Studies of Nucleic Acids   201

Contents

CHAPTER 8. **Lipids, Membranes, and Carbohydrates**

I.	Structures of Lipids and Membranes	208
II.	Raman Spectra of Lipids and Membranes	212
III.	Carbohydrates	232

Suggestions for Further Reading — 235

References — 238

Index — 253

Preface

In the past decade the application of Raman and resonance Raman spectroscopies to biochemical problems has opened up an exciting area of collaboration between the biochemist and the spectroscopist. Since a command of each other's field traditionally has not been essential, there are gaps in knowledge and communication that must be bridged in order to optimize interaction between the two areas of expertise. This book has grown from the need to have a readily available reference source that combines an introduction to both biochemical and spectroscopic terminology and theory. The aim is, on the one hand, to provide the biochemist with a basic grasp of Raman spectroscopy, explaining the information available from the technique and demonstrating its potential for biochemical research; and, at the same time, to introduce the spectroscopist to biochemistry, with particular reference to macromolecular systems and some of the questions that spearhead biochemical research. Explication of the main themes begins at an elementary level and assumes minimal familiarity with biochemical or spectroscopic jargon. The book sets out to make the novel and multidisciplinary aspects of the "bio-Raman" field available to chemists, biochemists, and spectroscopists at all stages of development from graduate student to experienced research worker.

The organization of the book naturally divides into two parts. The first deals with the spectroscopic aspects, while the second reviews biological systems and details the application of Raman spectroscopy to biological molecules. The first part is composed of Chapters 1, 2, and 3 and deals with the theoretical (at a nonmathematical level) and experimental aspects of Raman spectroscopy. The first chapter contains a very brief explanation of what Raman and resonance Raman spectroscopies are and a discussion of the advantages and disadvantages of the techniques for biochemical studies. The explanation of the Raman and resonance Raman effects is taken up in more detail in Chapter 2, and this is an important chapter for those with little or no background in vibrational spectroscopy. Chapter 2 develops the concept of the vibrational motions of molecules by initially considering mechanical "ball and spring" models and goes on to use this concept to formulate a classical model for Raman scattering. The resonance Raman effect is then described by another model which emphasizes the discrete or quantized energy levels available to a molecule. The experimental aspects of Raman spectroscopy are taken up in Chapter 3, which ranges from a description of the basic optics of a routine Raman experiment to a mention of some very sophisticated instrumental developments. The codevelopments of lasers, optics, and electronics continue to promote advances in experimental Raman spectroscopy, and these advances have important ramifications for biochemical applications. The second part of the book, which consists of Chapters 4–8, contains reviews of the application of Raman spectroscopy across the entire field of biochemistry. Chapters 4, 7, and 8 begin with an introduction to the three major classes of biological molecules, which are, respectively, proteins, nucleic acids, and lipids. To familiarize the nonbiochemist with these systems, each chapter contains an outline of the basic chemistry and biochemical nomenclature involved. The biochemical introductions are not extensive, but they should be enough to save the reader from constantly referring to a biochemistry text.

The process of researching and writing this book has been stimulated by the pleasure I have gained from the research of others. I have been constantly impressed by the audacity with which they have asked questions about some, at first sight, impossibly complex biological system and by how they have been rewarded with a remarkable answer about the workings of the molecular world. I hope some of my amazement and pleasure is transmitted by the following pages.

It is a pleasure to record my thanks to the authors and publishers who have allowed me to use their published work. In particular, I am grateful to Dr. R. H. Atalla and Dr. R. Mendelsohn, who offered some valuable comments on recent developments in the carbohydrate and the membrane fields, respectively. However, it goes without saying that I am solely responsible for the choice of material and its presentation. I am also indebted to Andy Storer for comments on part of the manuscript and to

Yukihiro Ozaki and Diana Pliura for reading the entire text and making many constructive suggestions. Finally, my thanks are due to Judi Meredith for typing the manuscript and to Celia Clyde and Denise Ladouceur for drawing and redrawing the figures; without their patient help this book could not have been written.

CHAPTER 1

Introduction

I. Raman Scattering

Raman spectroscopy is named after the Indian scientist C. V. Raman, who, with K. S. Krishnan, observed that some of the light scattered by a liquid is changed in wavelength (Raman and Krishnan, 1928). That portion of the light which undergoes a change in wavelength is known as *Raman scattered light*. The physical origin of Raman scattering lies in inelastic collisions between the molecules composing the liquid and *photons*, which are the particles of light composing the light beam. An inelastic collision means that there is an exchange of energy between the photon and the molecule with a consequent change in energy, and hence wavelength, of the photon (Fig. 1.1). Moreover, since total energy is conserved during the scattering process, the energy gained or lost by the photon must equal an energy change within the molecule. It follows that by measuring the energy gained or lost by the photon we can probe changes in molecular energy. The changes in the energy of the molecule are called *transitions* between *molecular energy levels*. In the biochemical context, Raman spectroscopy is concerned primarily with the vibrational energy level transitions of the molecule. In other words, by monitoring the inelastically scattered photons we can probe molecular vibrations, and *the Raman spectrum is a vibrational spectrum of a molecule*.

1

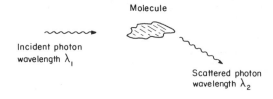

Fig. 1.1. An inelastic collision between a photon and a molecule.

As in the case of infrared spectroscopy, the utility of the Raman spectrum lies in the fact that the vibrational spectrum of a molecule is a sensitive indicator of chemical properties. The vibrational spectrum reflects the disposition of atomic nuclei and chemical bonds within a molecule and the interactions between the molecule and its immediate environment.

Raman and Krishnan used a beam of focused and filtered sunlight in their original observation of Raman scattering. Although modern Raman experiments use a laser as the light source, the technique remains conceptually simple and usually can be reduced to the schematic shown in Fig. 1.2. A monochromatic laser beam of wavelength λ_1 is focused into the sample to produce a high photon density, and the resulting scattered light is analyzed for wavelength and intensity. The Raman effect is extremely weak and only a minute proportion of the incident photons become Raman photons of wavelengths λ_2, λ_3, etc. The inherent weakness of the effect means that relatively high-power lasers must be used to create a high photon flux and that sophisticated optical and electronic equipment is required to detect the scattered photons.

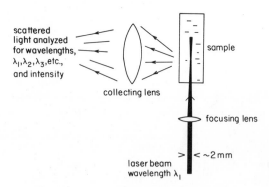

Fig. 1.2. Schematic of a Raman experiment.

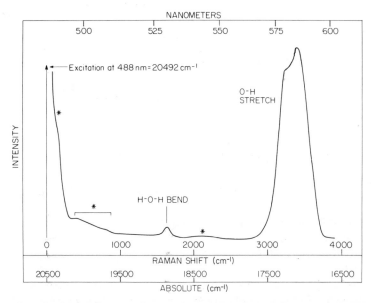

Fig. 1.3. A Raman spectrum of water. The broad features marked with an asterisk are due to interactions between water molecules which are not completely understood.

II. The Appearance of a Raman Spectrum

The appearance of a Raman spectrum is illustrated by the spectrum of water shown in Fig. 1.3. The most readily discernable peaks are the intense feature marked O–H stretch and the weak feature marked H–O–H bend. These Raman peaks arise, respectively, from an exchange of energy between the incoming photons and a vibrational motion corresponding to a stretching of the O–H bond or a bending motion of the water molecule.

Before we tackle the intricacies of the units given in Fig. 1.3, we can say immediately that the abscissa represents an energy difference between the scattered and incoming photons with the zero-point energy difference being at the wavelength of excitation (488 nm in Fig. 1.3). Therefore, it is apparent that more energy is required to bring about the O–H stretching motion than is required for the H–O–H bending motion. It is also apparent from the intensities of the peaks in the spectrum that the exchange of energy between molecules and photons, which gives rise to the O–H stretching peak is more probable than an exchange which gives rise to the H–O–H bending feature. Another important consideration in Fig. 1.3 is that, apart from the O–H stretching region, the Raman spectrum of water is weak and thus interferes only minimally with the Raman spectrum of a solute. This means that Raman spectroscopy can be carried out in water, the physiological solvent, and this is an important advantage for the Raman technique in biochemical studies. Further discussion of Fig. 1.3 requires definition of the basic units used in Raman spectroscopy.

III. Basic Units

A monochromatic light beam is characterized by its wavelength λ, its power, and its polarization. Here we shall be concerned with the relationships stemming from the wavelength of light and shall defer discussion of polarization and power until Chapter 2, Section VII and Chapter 3, Section I,A, respectively. Instead of quoting the wavelength of a light beam this property is often given in terms of the equivalent units of frequency or wavenumber. These quantities are related to wavelength thus:

$$\text{Frequency} = \nu = \frac{c}{\lambda} \quad \text{hertz (or cycles per second)} \qquad (1.1)$$

where c is the speed of light (2.998×10^{10} cm sec^{-1} in a vacuum) and λ is expressed in centimeters, and

$$\text{Wavenumber} = \frac{1}{\lambda} \quad \text{cm}^{-1} \qquad (1.2)$$

where λ is expressed in centimeters. Wavenumber is often denoted by ω or σ.

The wavelength of light is commonly expressed in ångströms (Å) or nanometers (nm)

$$10 \text{ Å} = 1 \text{ nm} = 10^{-9} \text{ meter} \qquad (1.3)$$

Finally, the wavelength (or frequency) of light is related to the energy E, by

$$E = h\nu \qquad (1.4)$$

Table 1.1

Color, Frequency, and Energy of Light

Color	Wavelength (nm)	Frequency (10^{14} Hz)	Wavenumber (cm^{-1})	Energy kJ/mole	Energy kcal/mole
Near infrared	1000	3.00	10,000	120	28.6
Red	700	4.28	14,300	171	40.8
Orange	620	4.84	16,100	193	46.1
Yellow	580	5.17	17,200	206	49.3
Green	530	5.66	18,900	226	53.9
Blue	470	6.38	21,300	254	60.8
Violet	420	7.14	23,800	285	68.1
Near ultraviolet	300	10.0	33,300	399	95.3
Far ultraviolet	200	15.0	50,000	598	143

where h is Planck's constant. By using Eqs. (1.1)–(1.4) we can set out the basic relationship between the wavelength, energy, and, incidently, color of light encountered in Raman experiments. This information is given in Table 1.1.

Having defined the important units, we can now return to Fig. 1.3 and explain the units given with that spectrum. The wavelength of the laser line used to excite the Raman spectrum is 488 nm, which is equivalent to 20,492 cm^{-1}. When discussing energies, it is simpler to use wavenumbers or frequencies since, from Eq. (1.4), energy is linear with wavenumber or frequency (and therefore inversely proportional to wavelength). The exchange of energy between the incoming photons and the vibrational energy transitions giving rise to the H–O–H bending and O–H stretching features results in a loss in energy for the photons equivalent to 1640 and 3400 cm^{-1}, respectively. Thus, on the scale of absolute cm^{-1}, the H—O—H bending and O–H stretch appear at 20,492 − 1640 = 18,852 cm^{-1} and 20,492 − 3400 = 17,092 cm^{-1}, respectively, and this can be seen by referring to the bottom scale in Fig. 1.3. The convention in Raman spectroscopy is to quote the positions of the vibrational peaks as the difference between the absolute wavenumbers of the exciting line and the absolute wavenumbers of the scattered photon. The Raman shift scale shown in Fig. 1.3 is normally the only scale encountered in published Raman spectra. The equivalent wavelength scale in nanometers is shown at the top of Fig. 1.3. As mentioned previously, the nonlinearity of the top scale reflects the fact that the spectrum is plotted as a linear function of energy but that energy is inversely proportional to wavelength.

IV. Raman, Resonance Raman, and Infrared
Spectroscopies

Raman spectra give detailed information on the vibrational motions of atoms in molecules, and, because these vibrations are sensitive to chemical changes, the vibrational spectrum can be used as a monitor of molecular chemistry. Essentially, the same kind of vibrational motion that produces the peaks in Raman spectra also produces the "peaks" (actually, absorption maxima) in infrared spectra. Hence much of the basic information that has been gained from infrared studies can be transferred to Raman spectroscopy and vice versa. However, the physics of the Raman effect and the infrared absorption process are quite different, and these dissimilarities have important implications for the applications of vibrational spectroscopy to an area such as biochemistry. The Raman effect is a *scattering process* in which the interaction between the photon and the molecule occurs in a very short time and Raman peaks correspond to photons that have "bounced," inelastically, off molecules. In contrast, "peaks" in an infrared spectrum correspond to energies where infrared photons have been *absorbed* by the molecule. This topic is discussed in more detail in Chapter 2, Section IV, but it should be stated here that for a complete description of the vibrational motion of a molecule both Raman and infrared spectra are needed. However, it is often impossible to obtain the infrared spectrum of a biological material in water; under these conditions, the Raman data are the sole source of information.

A major difference between the Raman and infrared techniques is that there is no infrared analog of the resonance Raman (RR) effect. The latter occurs when the laser wavelength, used to excite the Raman spectrum, lies under an intense electronic absorption band of a chromophore. Under these conditions considerable intensity enhancement of certain Raman bands may occur with the result that the absolute intensities are increased by a factor of 10^3 to 10^5. The absorption spectrum of the chromophore N-methyl-o-nitroaniline is shown in Fig. 1.4 which also illustrates the conditions for resonance Raman intensity enhancement. By excitation at 647.1 nm, a normal Raman spectrum of N-methyl-o-nitroaniline is obtained. However, by using a laser line which lies under the N-methyl-o-nitroaniline's intense absorption band, e.g., the 441.6-nm line shown in Fig. 1.4, a *resonance* Raman spectrum of the chromophore is obtained. Certain bands in the Raman spectrum are enhanced in intensity by a factor of approximately 2000 upon changing wavelength from 647.1 to 441.6 nm. Thus in practice the RR spectrum of nitroanilines can be obtained at concentrations as dilute as 10^{-4}–10^{-5} M in aqueous solution. The

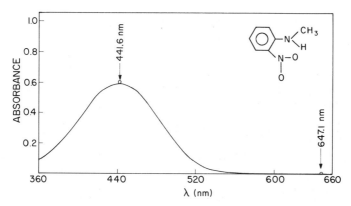

Fig. 1.4. The absorption spectrum of *N*-methyl-*o*-nitroaniline. A resonance Raman spectrum of this chromophore can be obtained by using an excitation wavelength lying within the absorption profile.

intensity enhancement associated with the resonance phenomenon introduces the possibility of *selectively* obtaining the vibrational spectrum of a single chromophore when the latter is found in a biological system consisting of many components. The nitroanilines can again be used to illustrate this point. Nitroanilines have been utilized to investigate the binding sites of certain protein molecules called antibodies (Chapter 6, Section III). Some nitroaniline derivatives bind strongly to antibodies. The absorption spectrum of *N,N*-dimethyl-2,4-dinitroaniline bound to an antibody pro-

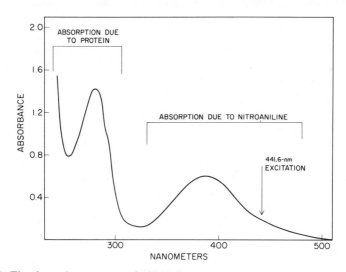

Fig. 1.5. The absorption spectrum of a *N, N*-dimethyl-2,4-dinitroaniline–antibody complex.

tein is shown in Fig. 1.5. The important observation from Fig. 1.5 is that the absorption maxima of the protein and of the nitroaniline are far apart. Thus, by using the 441.6-nm wavelength for excitation, it is possible to obtain a resonance Raman spectrum of the nitroaniline while a normal Raman spectrum of the protein is recorded. Since the RR spectrum is of high relative intensity, the measured spectrum contains features due to the nitroaniline bound to the antibody combining site while the weak features due to the normal Raman spectrum of the protein cannot be detected above the spectral noise level. Thus, in this example, the vibrational spectrum of the nitroaniline, of approximate molecular weight (MW) 200, can be selectively monitored when this molecule is bound to an antibody protein of MW 150,000.

V. Advantages and Disadvantages of Raman Spectroscopy for Biochemical Studies

Most of the advantages and disadvantages of using Raman spectroscopy in biochemical investigations are the same whether one is operating under Raman or resonance Raman conditions. The advantages and disadvantages are listed below.

A. Advantages

1. Raman spectra may be obtained for molecules in aqueous solutions since water has a weak Raman spectrum that interferes only minimally with the spectrum of the solute. In contrast, water has a very strong infrared absorbance which greatly hinders infrared (IR) spectroscopy of aqueous solutions.

2. Liquids, solutions, gases, films, surfaces, fibers, solids, and single crystals are equally amenable to examination.

3. Small amounts of material are required; the "active volume" of Raman scattering is governed by the size of the focused laser beam, typically 0.005×0.2 cm, and the sample need only fill this volume.

4. Raman spectroscopy permits diversity of experimental conditions: the maneuverability of the pencillike laser beam, in combination with advantages 2 and 3, allows great flexibility in experimental arrangements. Apparatus used in other areas of spectroscopy and biophysics is easily adapted for Raman use.

5. Homopolar bonds, such as C–C and S–S, give rise to intense or moderately intense Raman bands. In infrared spectra these bonds give rise to either weak or undetectable bands.

6. The entire vibrational spectrum from 10 to 4000 cm^{-1} can be covered in a single scan of a conventional Raman spectrometer. Again, this is an advantage of Raman over infrared spectroscopy since in the latter special instrumentation is normally required to cover the region below approximately 600 cm^{-1}.

7. The time scales of the Raman and resonance Raman effects are essentially instantaneous. Therefore, the Raman spectrum represents an instant "snapshot" of all molecules. Hence, in a system where rapid chemical exchange is occurring, each species contributes a Raman signature in direct proportion to its concentration. In general, relaxation effects on line shapes, such as those which are commonplace in nuclear magnetic resonance, are not readily observed.

8. Complete Raman spectra may be recorded in much less than 1 sec (Chapter 3, Section III,C).

B. Disadvantages

1. The high photon flux of the laser beam can produce unwanted photochemical effects.

2. The Raman, and even the RR, event is an improbable physical process, and the Raman spectrum is easily obscured by competing processes such as fluorescence. Experimental means are available to overcome the problems outlined in both disadvantage 1 and 2, and these are mentioned in Chapter 3.

3. Normal Raman spectroscopy requires concentrations considered by biochemists to be rather high, e.g., 0.1–0.01 M.

4. Normal Raman spectroscopy of solutions usually demands a high level of optical homogeneity.

C. Advantages Encountered Only in RR Spectroscopy

1. Concentrations of chromophore in the 10^{-4}–10^{-6} M range in H_2O may be used. Moreover, optical homogeneity is usually a less stringent requirement for RR experiments.

2. The intensity enhancement associated with the RR effect allows one to be selective and to "pick out" the vibrational spectrum of the chromophore when the latter is just one component in an extremely complex biological situation. Technical innovation is constantly extending the spectral range wherein resonance Raman spectroscopy can be practiced. Biological chromophores in the 250–750-nm range have been subjects of RR study.

3. The positions of peaks in Raman and RR spectra are a property solely of the electronic ground state. However, the intensity behavior of peaks in RR spectra is a source of information regarding the electronic excited states. This is a particular advantage in photobiological studies since in molecules such as chlorophyll and rhodopsin excited states are used to bring about chemical reactions. Excited electronic states of chlorophyll are pivotal in initiating photosynthesis, while the excited state of rhodopsin, formed by absorption of a photon, triggers a sequence of events leading to the visual process.

CHAPTER 2

Principles of Raman Spectroscopy

I. Introduction

The objective of this chapter is to introduce the concepts and terminology of vibrational spectroscopy to those who have little or no background in spectroscopy. Apart from a few equations in the last section only the most elementary mathematics is needed to follow the text. By mastering this chapter even the novice should be able to follow the rest of the book and to appreciate the scientific literature dealing with the biochemical applications of Raman spectroscopy.

II. The Vibrations of a Diatomic Molecule from a Classical Standpoint

Before we consider a mechanical model for a diatomic molecule, some fundamental concepts concerning vibrating systems are best illustrated by a single particle attached to a spring. This system is shown in Fig. 2.1, where the spring is fixed to a wall and is assumed to have no mass; the effect of gravity is ignored. Another important assumption at this stage is that the system obeys Hooke's law, which states that a particle, attached to a spring, when removed a distance from its equilibrium position, ex-

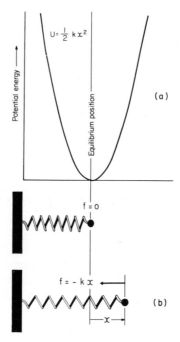

Fig. 2.1. (a) Potential energy function and (b) restoring force for a ball and spring system obeying Hooke's law. Adapted from G. M. Barrow, "Introduction to Molecular Spectroscopy," McGraw-Hill, 1962, by permission.

periences a restoring force that is proportional to its displacement from the equilibrium position. This can be written

$$f \propto x \qquad (2.1)$$

or

$$f = -kx \qquad (2.2)$$

where x, the measure of the distance from the equilibrium position, is the *displacement coordinate;* f is the force the spring imposes on the particle; and k is the proportionality constant called the *force constant*. The latter, whose counterpart occupies a very important place in actual molecular problems, measures the stiffness of the spring and thus gives the restoring force for unit displacement of the spring. The minus appears in Eq. (2.2) because as x increases in one direction the force is applied in the opposite direction.

The relationship between Hooke's law and potential energy must now be explored. The work required to displace the particle a distance dx is $f_{appl}\, dx$, and this work is stored as potential energy U. Therefore

$$dU = f_{appl}\, dx \tag{2.3}$$

Now,

$$f_{appl} = -f \tag{2.4}$$

since the applied force must be equal and opposite to the force exerted on the particle by the spring. Thus,

$$dU = -f\, dx \tag{2.5}$$

and

$$dU/dx = -f \tag{2.6}$$

After inserting Hooke's law [Eq. (2.2)] into Eq. (2.6) we have

$$dU/dx = kx \tag{2.7}$$

By integrating Eq. (2.7) and taking the equilibrium position as that of zero-point energy we find

$$U = \tfrac{1}{2}kx^2 \tag{2.8}$$

This potential function is shown in Fig. 2.1. The potential energy increases parabolically as the particle moves in either direction from the equilibrium position. When the parabolic potential function is used to approximate the potential of a chemical bond, the system is said to be treated as a *harmonic oscillator.* In dealing with the vibrations of complex molecules the assumption of potential energy functions analogous to Eq. (2.8) leads to an elegant and powerful means of analyzing the vibrations of such molecules. This topic is discussed more fully in Section VI.

Having introduced the concept of a force constant and having derived an equation for the potential energy of a vibrating particle that obeys Hooke's law, we can now consider the two-particle system shown in Fig. 2.2. The masses m_1 and m_2 are connected by a weightless spring and the particles are, for simplicity, allowed to move only along the line of the system. Obviously, this arrangement of "balls and a spring" provides the counterpart of the diatomic molecule problem. Using the equations of classical mechanics we shall investigate the nature of the motions of the particles shown in Fig. 2.2.

If r_1 and r_2 represent the displacements of the particles of mass m_1 and m_2, respectively, from the initial positions at which the particles were

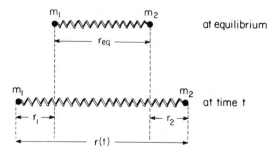

Fig. 2.2. Two particles connected by a "weightless" spring—the mechanical analog of a nonrotating diatomic molecule.

separated by their equilibrium distance and if Hooke's law is assumed for the spring, the kinetic and potential energies can be written

$$T = \tfrac{1}{2}(m_1\dot{r}_1^2 + m_2\dot{r}_2^2) \tag{2.9}$$

and

$$U = \tfrac{1}{2}k(r_2 - r_1)^2 \tag{2.10}$$

respectively, where

$$\dot{r} = dr/dt \tag{2.11}$$

By using Lagrange's equation, which for each particle is

$$\frac{d}{dt}\left[\frac{\partial T}{\partial r_i}\right] + \frac{\partial U}{\partial r_i} = 0 \tag{2.12}$$

we obtain two equations

$$m_1\ddot{r}_1 - k(r_2 - r_1) = 0 \tag{2.13}$$

and

$$m_2\ddot{r}_2 + k(r_2 - r_1) = 0 \tag{2.14}$$

which are statements of the generalization, "mass times acceleration $(m\ddot{r}_i)$ minus force $[k(r_2 - r_1)]$ equals zero." For differential equations of this form it is appropriate to look for solutions of the kind

$$r_1 = A_1 \cos(2\pi\nu t + \phi) \tag{2.15}$$

and

$$r_2 = A_2 \cos(2\pi\nu t + \phi) \tag{2.16}$$

These equations for r_1 and r_2 describe simple harmonic motions for both masses, each oscillating as a cosine function of time t, and it can now be seen how potential functions of the type in Eq. (2.8) give rise to harmonic oscillators. The oscillations occur with the same frequency ν (in hertz) and phase factor ϕ for both particles. However, the maximum amplitudes of oscillation A_1 and A_2 for each particle along the axis of the spring are different. We now substitute the Eqs. (2.15) and (2.16) into Eqs. (2.13) and (2.14) to find values for ν and A_1/A_2. By making these substitutions and rearranging we have

$$(-4\pi^2\nu^2 m_1 + k)A_1 - kA_2 = 0 \qquad (2.17)$$

and

$$-kA_1 + (-4\pi^2\nu^2 m_2 + k)A_2 = 0 \qquad (2.18)$$

From Eq. (2.17)

$$A_1/A_2 = k/(-4\pi^2\nu^2 m_1 + k) \qquad (2.19)$$

but from Eq. (2.18)

$$A_1/A_2 = (-4\pi^2\nu^2 m_2 + k)/k \qquad (2.20)$$

Therefore

$$k/(-4\pi^2\nu^2 m_1 + k) = (-4\pi^2\nu^2 m_2 + k)/k \qquad (2.21)$$

and by rewriting this relation

$$(4\pi^2\nu^2)^2 m_1 m_2 - 4\pi^2\nu^2 k(m_1 + m_2) + k^2 - k^2 = 0 \qquad (2.22)$$

we obtain the solutions

$$\nu = 0 \qquad (2.23)$$

and

$$\boxed{\nu = \frac{1}{2\pi}\sqrt{\frac{k}{\mu}}} \qquad (2.24)$$

where

$$\mu = m_1 m_2/(m_1 + m_2) \qquad (2.25)$$

The quantity μ is known as the *reduced mass*. Equation (2.24) is an important relation since, when we consider molecular vibrations, it can give an "intuitive feel" for the magnitude of vibrational frequencies and how these frequencies respond to molecular perturbations. It is apparent from Eqs. (2.23) and (2.24) that there are two natural frequencies for the system with one of the frequencies being zero.

The motions that correspond to these frequencies can be recognized if the relation of A_1 to A_2 is found.

Setting $\nu = 0$ in Eq. (2.17) or (2.18) gives

$$A_1 = A_2 \tag{2.26}$$

and therefore from Eqs. (2.15) and (2.16) $r_1 = r_2$. Thus, for $\nu = 0$ the particles move by the same amount and in the same direction, and the $\nu = 0$ solution corresponds to a translational motion of the entire system.

However, setting $\nu = (1/2\pi) \sqrt{k/\mu}$ into either Eq. (2.17) or Eq. (2.18) leads to

$$A_1/A_2 = -m_2/m_1 \tag{2.27}$$

and from Eqs. (2.15) and (2.16) at any time t

$$r_1/r_2 = -m_2/m_1 \tag{2.28}$$

Therefore the ratio of the amplitudes and the positions at time t is in inverse ratio to the masses. So, if the particles are of unequal mass, the lighter one will move with a greater amplitude than the heavier—in accord with a picture based on common sense. Both particles move with the same frequency $\nu = (1/2\pi) \sqrt{k/\mu}$, and both pass through their equilibrium positions at the same instant.

The choice of the coordinates used to describe the position of m_1 and m_2 is arbitrary. Previously, we used the displacements r_1 and r_2 since these are easy to visualize. However, by a suitable choice of coordinates the mathematical complexity involved in solving the equations of motion can be considerably reduced. In this instance the *internal coordinate*

$$q = r_2 - r_1 \tag{2.29}$$

is a useful choice since together with the *center-of-mass coordinate*

$$p = (m_1 r_1 + m_2 r_2)/(m_1 + m_2) \tag{2.30}$$

it may be shown that the equation of motion involving q simplifies to

$$\mu \ddot{q} + kq = 0 \tag{2.31}$$

with a solution of the form

$$q = A \, \cos(2\pi \nu t + \phi) \tag{2.32}$$

Equation (2.32) contains no contribution from p and should be compared to the two simultaneous equations (2.13) and (2.14), which were derived using r_1 and r_2 coordinates. The simplification arises from the fact that the choice of coordinates q and p eliminates the translational component of the problem (the $\nu = 0$ solution) before the vibrational aspect is developed.

III. A Classical Model for Raman Scattering

The Raman effect is a molecular light-scattering phenomenon in which a change in frequency of the light occurs. A light wave is a traveling wave of electric and magnetic fields, of which only the electric component produces Raman scattering. When a light wave meets a molecule composed of electrons and nuclei, the electric field of the wave at any instant is the same throughout the molecule because the molecule, perhaps 1 nm in size, is small compared to the wavelength of the light, which is typically 500 nm. Thus, the field exerts the same force on all electrons in the molecule and tends to displace them from their average positions around the positively charged nuclei. It is crucial for the Raman process that the displacements result in an *induced dipole moment* $\boldsymbol{\pi}$ in the molecule which is, to a good approximation, proportional to the electric field strength E. Thus

$$\boldsymbol{\pi} = \boldsymbol{\alpha}\mathbf{E} \tag{2.33}$$

where the proportionality factor $\boldsymbol{\alpha}$ is called the *electric polarizability* of the molecule. In general, the vector $\boldsymbol{\pi}$ has a different direction from that of the vector $\mathbf{E}$, and therefore $\boldsymbol{\alpha}$ is not a simple scalar quantity. In fact, the magnitudes of the three components defining $\boldsymbol{\pi}$, namely, π_x, π_y, and π_z, are related to the magnitudes of the electric field E by the three relations

$$\pi_x = \alpha_{xx}E_x + \alpha_{xy}E_y + \alpha_{xz}E_z$$
$$\pi_y = \alpha_{yx}E_x + \alpha_{yy}E_y + \alpha_{yz}E_z \tag{2.34}$$
$$\pi_z = \alpha_{zx}E_x + \alpha_{zy}E_y + \alpha_{zz}E_z$$

These equations express the fact that all three components of E contribute to each of three components of $\boldsymbol{\pi}$. The nine coefficients α_{ij} are called the components of the polarizability $\boldsymbol{\alpha}$ with the subscript i denoting the component of $\boldsymbol{\pi}$ and the subscript j denoting the component of E related by the element α_{ij}. In consequence of Eqs. (2.34), $\boldsymbol{\alpha}$ is said to be a tensor.

The electric field of the light wave varies with time. If a fixed molecule is irradiated with monochromatic radiation of frequency ν_0, expressed in hertz (cycles per second), which is plane polarized in the z direction, E_z as a function of time is given by

$$E_z(t) = E_{\max} \cos 2\pi\nu_0 t \tag{2.35}$$

where E_{max} is the value of E_z at its maximum and t is the time in seconds from an arbitrary starting time. Thus, for the z component of π

$$\pi_z(t) = \alpha_{zz}E_{max} \cos 2\pi\nu_0 t \qquad (2.36)$$

Since π_z depends on α_{zz} as well as E_z the properties of the molecule can change π_z via $\boldsymbol{\alpha}$. In the present context $\boldsymbol{\alpha}$ varies with time as a consequence of the vibrations of the molecule since the ease with which electrons may be displaced by the electric field depends on how tightly they are bound to the nuclei, which in turn depends on the internuclear separation. The result of the time dependence of $\boldsymbol{\alpha}$ and E on π can be seen by considering the simple example of a single diatomic molecule with nuclei of mass m_1 and m_2 (Fig. 2.2).

We saw in Section II (Eq. 2.32) that by choosing a suitable internal coordinate q, which represents the difference in displacements $r_2 - r_1$ of the nuclei m_1 and m_2 from their equilibrium positions, the dependence of q on time could be written

$$q(t) = A \cos(2\pi\nu_{vib}t + \phi) \qquad (2.37)$$

where ν_{vib} is the vibrational frequency of the molecule in hertz. This can be stated in simpler terms conceptually if we recognize that $q = r_2 - r_1 = \Delta r(t)$, which is the difference from equilibrium in the internuclear distance at time t (see Fig. 2.2). Thus

$$\Delta r(t) = r(t) - r_{equil} \qquad (2.38)$$

Furthermore, the amplitude A can be equated to the maximum extension of the distance between the two nuclei Δr_{max}. Equation (2.37) can then be written

$$\Delta r(t) = \Delta r_{max} \cos(2\pi\nu_{vib}t + \phi) \qquad (2.39)$$

Finally, by taking the time t to have the same starting point as the time scale for the light wave in Eq. (2.35), the phase constant ϕ may be equated to zero.

Using the postulate that the polarizability of the diatomic molecule depends linearly on Δr we can now write for the α_{zz} component of α

$$\alpha_{zz}(t) = \alpha_{zz}^{equil} + (d\alpha_{zz}/dr)\,\Delta r(t)$$
$$= \alpha_{zz}^{equil} + (d\alpha_{zz}/dr)\,\Delta r_{max} \cos 2\pi\nu_{vib}t \qquad (2.40)$$

The constant α_{zz}^{equil} is the polarizability element of the nonvibrating molecule and $d\alpha_{zz}/dr$ characterizes the manner in which the polarizability changes with r.

After substituting the expressions for α_{zz} into Eq. (2.36) we have the dependence of π on the time fluctuations of both α and E:

$$\pi_z(t) = \alpha_{zz}^{\text{equil}} E_{\text{max}} \cos 2\pi\nu_0 t$$

$$+ (d\alpha_{zz}/dr) \, \Delta r_{\text{max}} E_{\text{max}} \cos 2\pi\nu_0 t \cos 2\pi\nu_{\text{vib}} t \qquad (2.41)$$

By using the identity $\cos\theta \cos\phi = \frac{1}{2}[\cos(\theta + \phi) + \cos(\theta - \phi)]$ we can write Eq. (2.41)

$$\pi_z(t) = \alpha_{zz}^{\text{equil}} E_{\text{max}} \cos 2\pi\nu_0 t + \frac{1}{2}(d\alpha_{zz}/dr) \, \Delta r_{\text{max}} E_{\text{max}} \cos 2\pi(\nu_0 + \nu_{\text{vib}})t$$

$$+ \frac{1}{2}(d\alpha_{zz}/dr) \, \Delta r_{\text{max}} E_{\text{max}} \cos 2\pi(\nu_0 - \nu_{\text{vib}})t \qquad (2.42)$$

Equation (2.42) demonstrates that when a light wave interacts with a vibrating diatomic molecule, the induced dipole moment, in this case exemplified by π_z, has three components contributing to its time dependence. The first term on the right-hand side of Eq. (2.42) is a component vibrating with the frequency of the incident light and with a magnitude determined by $\alpha_{zz}^{\text{equil}}$ and E_{max}. According to classical electromagnetic theory, an oscillating dipole radiates energy in the form of scattered light. Thus, as a result of the first term in Eq. (2.42), light of the incident frequency ν_0 will be emitted and will be observable in directions which differ from that of the incoming light. This is the phenomenon of *Rayleigh scattering.* The second term is a component vibrating at a frequency which is the sum of the frequencies of the light and the molecular vibration. The scattered light arising from this second term is known as *anti-Stokes Raman scattering.* The third term is a component vibrating at a frequency given by that of the light wave minus that of the molecule and the scattered light resulting from this term is known as *Stokes Raman scattering.* Both these components have magnitudes which depend on the field strength of the light, the amplitude of vibration, and the polarizability derivative $d\alpha_{zz}/dr$. The appearance of scattered radiation, e.g., at $\nu_0 + \nu_{\text{vib}}$, in units of hertz arising from the second term in Eq. (2.42), or $\nu_0 - \nu_{\text{vib}}$ from the third term in that equation, means that by analyzing the scattered light we can monitor the vibrations within a molecule. It is this ability to measure molecular vibrations which gives the Raman effect its importance in the study of molecules in general and, in the present context, in the study of biochemical systems.

IV. Comparison with Infrared Spectroscopy

The information about molecular vibrational frequencies provided by infrared absorption spectroscopy is of the same kind as that provided by the ν_{vib}s of the Raman lines. Moreover, for molecules possessing little or

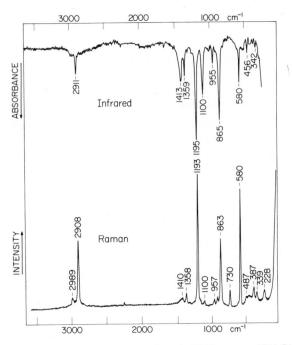

Fig. 2.3. The infrared and Raman spectra of methyldithioacetate [$CH_3C(=S)SCH_3$] in the liquid phase.

no symmetry the infrared and Raman spectra can have quite similar appearances. This similarity is shown in Fig. 2.3, which compares, on the same wavenumber scale, the Raman spectrum of methyldithioacetate ($CH_3C(=S)SCH_3$) with the infrared absorbance spectrum. The "peaks" in the infrared spectrum, where maximum absorption of the infrared radiation occurs, frequently coincide with "peaks" of maximum intensity of Raman scattering. However, this is not always the case, e.g., the 1100-cm^{-1} infrared feature can only just be detected in the Raman spectrum. Hence both techniques are needed to form a complete picture of the vibrational behavior of a molecule. Information from both techniques is especially important for molecules possessing a center of symmetry since there can be almost exact complementarity between the Raman and infrared spectra with no correspondence between the main features exhibited in each spectrum. The complementarity derives from the *rule of mutual exclusion,* which states that for molecules with a center of symmetry, vibrational transitions that are allowed in the infrared are forbidden in the Raman effect and vice versa.

The reason for the complementarity of Raman and infrared spectra lies in the different natures of the physical processes involved in the two effects. As outlined in the previous section, the Raman process is a scattering effect involving an *induced* dipole π, which, in turn, depends on a change in molecular polarizability during a vibration. For the diatomic molecule, with nuclei m_1 and m_2, discussed in Section III, the change in polarizability is the $d\alpha_{zz}/dr$ term that appears in Eq. (2.42). In contrast, infrared spectroscopy is an absorption process caused by a change in the *permanent* molecular dipole μ with change in bond length during a vibration, e.g., in the diatomic molecule m_1m_2, $d\mu/dr$ leads to the absorption of radiation frequency $\nu_{\rm vib}$ which is characteristic of the bond linking m_1 to m_2. Since frequencies of the kind $\nu_{\rm vib}$ correspond to the characteristic frequencies in the mid and far infrared regions, this absorption process occurs in the infrared region of the spectrum.

The relationships defining the intensities of the Raman lines and infrared bands are quite different for the two spectra. Transmission of a collimated monochromatic beam of infrared radiation of frequency ν depends exponentially on the sample's thickness x and its absorptivity a_ν:

$$\%T_\nu = 100(I/I_0)_\nu = 100 \exp(-a_\nu x) \tag{2.43}$$

where I_0 and I are the intensities of the beam before and after transmission through the sample. The absorptivity a_ν, is determined by the square of the first derivative of the dipole moment with respect to the normal coordinate (i.e., the $d\mu/dr$ term for the simple diatomic). The relationship for Raman intensities is far more complex and is considered in Section VIII; however, it can be said at this point that the intensity of Raman scattering is directly proportional to the intensity of the exciting radiation and the number of molecules irradiated.

V. A Quantum Mechanical Picture
of Raman Scattering

The classical model discussed in Section III provides a useful conceptual model for Raman scattering. The quantum mechanical approach to the scattering process is quite different: the wave–particle duality of a light beam is incorporated by considering that the beam is composed of packets or quanta of light particles known as photons. Moreover, the quantization of molecular energy levels is taken into account, and a means is provided for calculating the polarizability α, and thus Raman intensities, in terms of the electronic properties of a molecule. In this section we are concerned with a model built around quantized vibrational and

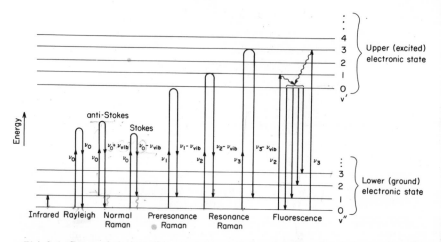

Fig. 2.4. Some of the possible consequences of a photon–molecule interaction. The lengths of the upward-pointing arrows are proportional to the frequencies of the incoming light while the lengths of the downward-pointing arrows are proportional to the frequency of the scattered (or in the case of fluorescence, emitted) light. The vibrational quantum numbers in the upper and lower electronic states are v' and v'' respectively. The energy spacing v'' between the lower state vibrational levels is equal to ν_{vib}.

electronic energies; we leave the subject of calculating α and Raman intensities until Section VIII.

It is a reasonable approximation, for a gaseous diatomic molecule, to write the molecular energy E_{mol} as a sum of terms

$$E_{\mathrm{mol}} = E_{\mathrm{elec}} + E_{\mathrm{vib}} \qquad (2.44)$$

where the subscripts refer to electronic and vibrational components, respectively of the total energy. For present purposes the contributions due to molecular rotation and translation can be ignored. Electronic energy transitions involve much larger quantities of energy than vibrational transitions do, with values of 10,000–50,000 cm^{-1} for the former and 10–4000 cm^{-1} for the latter (these numbers convert to 28.6–143 and 0.028–11.4 kcal/mole of quanta, respectively). This situation is depicted in Fig. 2.4 with a large energy spacing between the ground and excited electronic states, and smaller spacings between the vibrational levels contained within each state. Moreover, a good approximation for the vibrational energy E_{vib} of a diatomic molecule is

$$E_{\mathrm{vib}} = (v + \tfrac{1}{2})\nu_{\mathrm{vib}}, \qquad v = 0, 1, 2, \ldots \qquad (2.45)$$

where the vibrational quantum number v has only integral values so that the vibrational energy levels in the ground state in Fig. 2.4 are equally spaced by the amount ν_{vib}. The energy E_{vib}, of course, takes the units of ν_{vib} (hertz if the latter is expressed in hertz and so on).

In the quantum mechanical model, light scattering is depicted as a two-photon process. The first step in this process is the combination of a photon and a molecule to raise the molecule to a higher energy state of extremely short lifetime. This transition is depicted by the upward-pointing arrows in Fig. 2.4, and as shown in the figure, the higher energy state may or may not correspond to a quantized energy state of the molecule. The second step, indicated by the downward-pointing arrows in Fig. 2.4, involves the release of a photon after a very short time interval ($<10^{-11}$ sec). The energy of this second photon is given by the length of the downward-pointing arrows in Fig. 2.4. For Rayleigh scattering the upward and downward directed transitions have the same length and have, therefore, apart from a change in sign, the same energies. Thus, in the Rayleigh process, no change in frequency of the photon occurs.

The various kinds of Raman processes may now be outlined. If the downward-pointing arrow stops on a vibrational energy level that is higher than the starting level, a Stokes process has occurred. In this, the second photon has a frequency $\nu_0 - \nu_{vib}$ corresponding to the third term in Eq. (2.42). Conversely, an anti-Stokes process results from the transition terminating at a vibrational energy level lower than the starting level. In the anti-Stokes process, the second photon has an energy $\nu_0 + \nu_{vib}$, giving the same result as the second term in Eq. (2.42). Of course, in both processes total energy is conserved; thus for Stokes scattering the molecule gains a quantum of energy ν_{vib} while for anti-Stokes scattering the reverse is true. For both Stokes and anti-Stokes processes a selection rule can be derived from Eq. (2.45) which says that v'', in Fig. 2.4, can only change by ± 1. Thus, Eq. (2.42), derived from the classical model, agrees with the results obtained by considering quantized energy levels; both models predict that the difference in frequency between the incident and scattered light $\nu_0 - (\nu_0 - \nu_{vib})$ corresponds directly to the molecular vibrational frequency ν_{vib}. The quantum mechanical model also illustrates a very important generalization, namely, *that the position of Raman peaks is a property solely of the electronic ground state*. This follows from the fact that ν_{vib} is a vibrational transition within the lower or ground electronic state in Fig. 2.4.

In the classical model, Eq. (2.42) indicates no difference in the expected intensities of Stokes and anti-Stokes transitions since the coefficients of the two terms in the equation are the same. However, the model of quantized energy levels depicted in Fig. 2.4 shows that for anti-Stokes transitions to take place the molecule must be in a higher ($v'' > 0$) vibrational state within the electronic ground state. Since the population of these higher states is governed by a Boltzmann distribution, only a small percentage of molecules are in higher vibrational states. The ratio of the number of molecules in the $v'' = 1$ and $v'' = 0$ vibrational states in the ground electronic state (Fig. 2.4), which we denote N_1 and N_0, respectively, is

$$N_1/N_0 = \exp(-h\nu_{\mathrm{vib}}/kT) \qquad (2.46)$$

where h is Planck's constant, T is the absolute temperature, and k is Boltzmann's constant. When $T = 300°K$ and the vibrational frequency is 480 cm^{-1}, N_1 is 0.1 N_0. As a result of the exponential nature of Eq. (2.46) an anti-Stokes line at 3×480 cm^{-1}, or 1440 cm^{-1}, would be 0.001 as strong as the corresponding Stokes line. In practice this means that the feeble anti-Stokes scattering is usually ignored in conventional Raman spectroscopy and only the Stokes spectrum is recorded.

Having outlined normal Raman scattering in which the energy of the incident light is considerably less than that needed to reach the higher energy electronic state in Fig. 2.4, we now consider the result of the light beam energy approaching that of the energy gap between the lower and higher electronic states. In Fig. 2.4 the transition labeled "preresonance Raman" is attributable to a light frequency that has almost enough energy to produce direct electronic absorption by the molecule. Under this condition the intensity of Raman scattering shows a marked increase. For normal or nonresonant conditions, Raman intensities are proportional to the fourth power of the *scattered* light frequency ν_s. However, as preresonance Raman conditions are approached, the intensity of scattering increases much more rapidly than ν_s^4.

A slight increase in the energy of the exciting radiation over that for the preresonance case places the upward transition in Fig. 2.4 within the higher electronic state. Absorption of a photon can now occur, and by the prompt reemission of a second photon can give rise to the resonance Raman process. As will be discussed in Section VIII, band intensities in resonance Raman spectra can be orders of magnitude greater than those in normal Raman spectra. Apart from making spectra intrinsically easier to detect, the resonance Raman effect introduces the possibility of selectively probing the vibrational spectrum of a small but key part of an extremely complex biological system. This is achieved by choosing a system that contains a chromophore of interest which absorbs photons in or near the visible spectral region. By exciting the Raman spectrum with light that can be selectively absorbed by the chromophore the resonance Raman spectrum of that chromophore may thereby be obtained uncluttered by the normal Raman spectrum from the rest of the system, which, being orders of magnitude less intense, is lost in the spectral background.

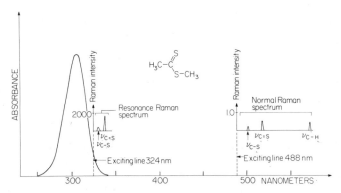

Fig. 2.5. The main features in the absorption, resonance Raman and normal Raman spectra of methyldithioacetate showing their relative positions on the wavelength scale. The choice of the 488-nm exciting line is arbitrary since any wavelength longer than 400 nm would yield a normal Raman spectrum.

These concepts can be understood by referring to Fig. 2.5, which depicts the electronic absorption spectrum of the dithioester chromophore, in this case contained in $CH_3C(=S)—SCH_3$. Two possible excitation lines for the observation of normal and resonance Raman spectra are indicated together with the position of the Raman spectrum in each case. The excitation line used to produce the resonance Raman spectrum should be near 310 nm to give the highest degree of intensity enhancement. The 324-nm line depicted in Fig. 2.5 was used simply because of its availability in the author's laboratory. The peaks due to stretching of the C—S and C=S bonds are intense in both the normal and resonance Raman spectra. However, the intensities of both peaks increase by a factor of 2000 on going from 488- to 324-nm excitation. The positions of the peaks, in wavenumbers from the exciting line, are the same in both spectra, but the resonance Raman spectrum appears "squashed" since the abscissa is not linear in energy. The ν_{C-H} feature due to the stretching motion of the C–H bonds does not appear under the conditions used to record the resonance Raman spectrum since the C–H bonds undergo little change in geometry in the excited compared to the ground electronic state; consequently, the C–H mode is not intensity enhanced (Section VIII).

The dithioester chromophore can also be generated in the active site of a protein of molecular weight of approximately 25,000, and by employing laser excitation in the 310–340-nm region, the resonance Raman spectrum of the chromophore is obtained free from spectral features due to the protein. In this way the resonance Raman spectrum provides a specific probe of a key site in a complex multithousand atom environment. Other examples can be found in many biochemical systems, and these are discussed mainly in Chapters 5 and 6.

The foregoing description of the resonance Raman interaction between a molecule and a photon should not be taken to imply that this is the only consequence of photon absorption. In fact, two other processes compete with the resonance Raman effect, and both of these occur with higher probability than resonance Raman scattering. In one process the molecule may not reemit any radiation at all following the absorption of a photon; in this case the molecule returns to the lower electronic state by a nonradiative pathway and dissipates the energy of the photon as heat. The second process is termed fluorescence and for large biological molecules in solution involves the emission of a photon from the $v' = 0$ vibrational state of the upper electronic state in Fig. 2.4. Fluorescence emission is slow compared to the resonance Raman effect with the former taking more than 10^{-9} sec whereas the time scale of the resonance Raman interaction is shorter than 10^{-11} sec. In the fluorescence process it is the time delay of more than 10^{-9} sec inherent in the effect that allows the molecule to lose any excited vibrational energy and degenerate into the $v' = 0$ state of the upper electronic level prior to emission of the fluorescence photon. From the practical standpoint this means that the downward-pointing arrows in the fluorescence process depicted in Fig. 2.4 are always the same length. Thus the absolute frequency of fluorescence photons is independent of the excitation frequency ν_0 in contrast to the absolute frequency of Raman photons, which is $\nu_0 - \nu_{vib}$ and therefore does depend on the excitation frequency.

VI. The Vibrations of a Polyatomic Molecule

The following discussion of the vibrations of a polyatomic molecule applies equally well to infrared and Raman spectroscopy; therefore examples are given of the use of both techniques in a biochemical context. It has already been mentioned that the two forms of vibrational spectroscopy are, in a sense, complementary since certain vibrations give rise to intense infrared absorption but only weak Raman scattering and vice versa. Unfortunately, it is often not possible to apply both techniques to the same biochemical system. The reasons why Raman spectroscopy is often the method of choice have been outlined in Chapter 1, Section V, and are, principally, that water has an intense infrared absorbance which can obliterate the infrared spectrum of any solute and that, using the resonance Raman effect, Raman spectroscopy can be applied to chromophores at low concentration. For example, the ability of RR spectroscopy to obtain the vibrational spectrum of the heme group in hemoglobin at 10^{-4}–10^{-5} M concentration cannot be matched by infrared techniques. However, in some systems (e.g., lipid dispersions; see Chapter 8, Section II) the Fourier transform infrared technique can provide data comparable to those gained from normal (nonresonance) Raman spectroscopy.

A. Degrees of Freedom of Motion

Since atomic nuclei are massive compared to electrons, it is a good approximation to describe the dynamics of a molecule in terms of the motions of the nuclei alone. Each nucleus requires three coordinates to describe its position (e.g., in a Cartesian coordinate system, x, y, and z). Thus each nucleus has three independent degrees of freedom, and for a molecule containing N atomic nuclei the total number of degrees of freedom is $3N$. In the present context, however, we are interested in only the vibrational dynamics of the molecule and we can subtract out the overall translational and rotational motions since these do not affect the internal vibrations. The center of gravity of the molecule requires three coordinates to define its position, and thus there are three independent degrees of freedom of motion associated solely with the translations of the center of gravity. Similarly, when a nonlinear molecule is in its equilibrium configuration, it requires three additional coordinates to describe its rotational motion about the center of gravity. A nonlinear molecule, therefore, has three independent rotational degrees of freedom. A linear molecule, however, has only two rotational degrees of freedom. This arises from the fact that rotation about the molecular axis is not considered a degree of freedom of motion since the nuclei are not displaced by such an operation. By subtracting the translational and rotational degrees of freedom from the original total of $3N$ we see that a nonlinear molecule has $3N - 6$ internal degrees of freedom and a linear molecule has $3N - 5$ internal degrees of freedom. Translations of the center of gravity and rotations about this point occur without changing the shape of the molecule. In contrast, the internal degrees of freedom, which are associated with vibrational motions, change the shape of the molecule without moving the center of gravity or rotating the molecule.

B. Normal Modes of Vibration and Normal Coordinates

If we were able to watch a movie depicting the vibrational motion of a complex molecule composed of N atoms, it would be impossible to discern any order or pattern to the vibrations. However, it can be shown that the complicated motions of any molecule can be broken down into simple components known as the *normal vibrations* (also called the *normal modes*). Furthermore, the $3N - 6$ internal degrees of freedom of a non-linear molecule correspond to $3N - 6$ independent normal modes of vibration. In each normal mode all the atoms in the molecule vibrate with the same frequency, and all the atoms pass through their equilibrium positions simultaneously. The resolution of the vibrational motions into simple normal modes depends on making an approximation for the potential energy similar to Eq. (2.8), namely, that the potential energy is proportional to the square of the displacement from the nuclear equilibrium position. This, of course, finds its equivalent in the "ball and spring" models as Hooke's law. The same assumption also leads to the result that in each normal mode every mass performs simple harmonic motion with the same characteristic frequency ν_{vib}.

As an example of the normal modes of a molecule of biological importance consider carbon dioxide. Since it is linear, CO_2, on the basis of the above, has $3 \times 3 - 5 = 4$ degrees of freedom and hence 4 normal modes. These are shown in Fig. 2.6. The filled circles in the figure represent the equilibrium positions of the nuclei, and the arrows represent the direction of their displacement at some particular time. The normal vibration ν_1 is termed the symmetrical stretching mode and ν_3 the antisymmetrical stretching mode. In addition, there are two bending modes ν_{2a} and ν_{2b}.

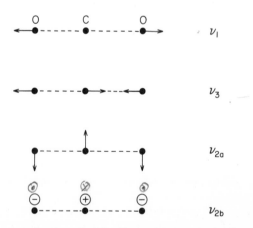

Fig. 2.6. The normal modes of carbon dioxide. The nuclei move in the direction of the arrows. In the ν_{2b} mode the $\oplus$ and $\ominus$ signs indicate that the nuclei move perpendicular to the plane of the page toward and away from the reader, respectively.

These have the same frequency, but since they take place in perpendicular planes (ν_{2a} is drawn in the plane of the paper and ν_{2b} perpendicular to it), either one may be stimulated without exciting the other. The bending modes are an example of *degenerate vibrations,* in this case they form a mutually degenerate pair.

We now consider coordinate systems. In our discussion of the diatomic molecule in Section II we worked through the equations of motion using Cartesian coordinates. However, it was pointed out at the end of Section II that considerable simplification in the mathematics could be brought about by choosing a coordinate system involving an internal coordinate (the bond length) and a center of mass coordinate. The value of this choice is that the translational motion of the molecule is "eliminated" at the start, leaving just the vibrational motions to be characterized.

Normal modes of vibration can be described using either Cartesian coordinates or internal coordinates involving change in bond lengths or bond angles. However, from the standpoint of analyzing the equations of motion, a third coordinate system involving *normal coordinates* has particular advantages. Normal coordinates are chosen so that each normal mode of vibration is characterized by a single coordinate called the normal coordinate Q. Q varies periodically with time, and as one normal coordinate vibrates, every Cartesian displacement coordinate and every internal coordinate vibrates, each with an amplitude in a specified proportion to the amplitude of the normal coordinate. The proportionality factors relating the amplitudes are such that the resulting motion is a normal mode. The mathematical advantage arises from the fact that each normal mode may be excited without affecting any of the other normal modes. Each normal coordinate makes a separate and independent contribution to the kinetic and potential energy terms used to derive the equations of motion outlined in Section II. This allows the contribution from each normal mode to be treated separately.

C. Group Frequencies

The interpretation of the Raman spectra of biochemical molecules in terms of their chemical properties is the key to using Raman spectroscopy in biochemistry. In this section we shall concentrate on relating the frequencies of Raman scattered light (i.e., the positions of peaks in a Raman spectrum) to molecular conformation and environment. The intensities and polarization properties of Raman peaks are additional valuable sources of molecular information, but these topics will be dealt with in Sections VIII and VII, respectively.

The positions of peaks in vibrational spectra may be analyzed by a relatively rigorous mathematical treatment along the lines of the equations of motion set out for a diatomic molecule in Section II; however, for polyatomic molecules the equations use normal coordinates to simplify the calculations. Both infrared and Raman data are needed for normal coordinate calculations and though the method directly calculates only peak positions, the analysis can be simplified by taking into account the symmetry properties of the molecule and the intensities and polarization of the vibrational bands. Moreover, it is necessary to assume a geometric model for the molecule by utilizing results obtained by X-ray diffraction or other spectroscopic techniques. For large asymmetric molecules, where the number of atoms is greater than, say, 20, normal coordinate calculations become impracticable. However, a wealth of information has been accumulated for small molecules, and from this it is possible to "carry over" certain generalizations to aid our understanding of large molecule spectra. For example, it has been shown that atoms tend to move either along chemical bonds or perpendicular to them, and this allows the classification of stretching and bending vibrations, respectively. Furthermore, since it is usually easier to bend a bond rather than to stretch it, frequencies of stretching vibrations are typically two or three times larger than those of bending vibrations. This is demonstrated by the spectrum of water shown in Fig. 1.3 where the O–H stretch occurs at a higher Raman shift compared to the H–O–H bend. Importantly, it is sometimes found that certain normal modes for a polyatomic molecule are dominated by the motions of just a few neighboring atoms. In this case the group of atoms has a characteristic vibrational frequency known as a group frequency.

The idea that certain chemical functional groups, such as the C—H, C=O, or C≡C moieties met in organic chemistry, give rise to characteristic vibrational frequencies first arose from experimental evidence dating back nearly 100 years. It was found, for example, that all molecules containing the —C—H group have a feature in their vibrational spectra near 2960 cm^{-1}, while if C—H is present as part of the C—H grouping of any molecule a mode always occurs with a frequency of 3020 cm^{-1}. Similarly, all molecules possessing ≡C—H have a vibrational mode of frequency 3300 cm^{-1}. Analogous correlations were found for many other organic and inorganic functional groups, and, on this basis, the presence of a characteristic frequency in the Raman or infrared spectrum could be taken as supportive evidence for a certain group in a molecule of unknown structure.

The existence of group frequencies is attributable to the fact that the force constant for a given chemical bond is, to a degree, transferable from one molecule to another. The validity of the concept also rests on the finding that some normal modes do correspond quite closely to the motion of just the functional group under consideration since, although in each normal mode of a polyatomic molecule every atom undergoes a periodic oscillation with the same frequency ν, the amplitudes of each oscillation are generally different. In a normal mode that gives rise to a useful group frequency, the atoms (or atom) within that functional group have large amplitudes of vibration, but the amplitudes of the other atomic oscillations are relatively small. Therefore, for the group frequency concept to apply to a part of a molecule, the motion in a given normal mode must be essentially localized within that group. This being the case, the group can be considered as an *"independent oscillator."*

Group frequencies find maximum utility when they give rise to intense features, when they occur in a spectral region that is free from other intense features, and when small variations in the group frequency can be correlated with conformational and environmental changes. The vibrations of the S–H, or C–H, or O–H, or N–H moieties within molecules are prime examples of the group frequency concept, and it is easy to see how characteristic frequencies arise from these linkages. Since the mass of the hydrogen nucleus is so much smaller than that of the other nuclei in the molecule, the amplitude of the vibrations of each hydrogen nucleus is much larger than those of any other nuclei. Thus, to a reasonable approximation, we may envisage the H nucleus oscillating against an infinitely large mass with the vibrational frequency of the hydrogen atom depending only on the force by which it is bound to the rest of the molecule. Consequently, the frequency will be the same for different molecules with the same S–H, or C–H, or O–H, or N–H force constants.

The N–H and O–H group frequencies have been used with great effect to study hydrogen bonding. When the hydrogens of either of these groups form a strong hydrogen bond to an electronegative atom, there is a characteristic and quite large shift in the X–H stretching mode to lower frequencies due to a reduction of the strength of the X–H bond on complexation and a concomitant drop in the force constant. There have been many studies on this topic which have helped to identify hydrogen bonds and to quantitate their behavior. In a qualitative biochemical context, the N–H stretching region in the infrared spectrum was used by Kyogoku *et al.* (1966) to show that specific association between the nucleic acid constituents guanosine and cytosine and between adenine and uracil occurs in chloroform solution. For example, Fig. 2.7 shows the N–H stretching region in the infrared spectra of guanosine and cytosine alone and guanosine and cytosine in the same solution. The strong perturbations to the N–H features in the observed spectrum of the mixture compared to

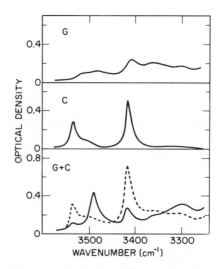

Fig. 2.7. Infrared spectra in the N–H stretching region for derivatives of guanosine (G) and cytidine (C). The calculated spectrum (broken curve) represents the sum of the upper two spectra obtained if the molecules did not interact. The observed spectrum (solid curve), taken at [G] = [C] = 0.0008 *M,* differs considerably from the calculated sum, showing that marked association is occurring. Adapted from Kyogoku *et al.* (1966). Copyright 1966 by the American Association for the Advancement of Science.

that of the calculated sum of the components show that specific guanosine–cytosine association involving N–H hydrogen bonds is occurring, although because of the number of N–H groups contributing to the spectrum an exact model could not be proposed for the structure of the complex.

An example of how group frequencies may be useful in even the most complex molecules centers around the normal modes of the amide group. All amides, polypeptides, and proteins (Chapter 4) show intense features in their infrared and Raman spectra near 1660 cm^{-1}, which are attributable to the so-called amide I mode. For *N*-methylacetamide, Miyazawa *et al.* (1958) showed that the amide I vibration corresponded to the normal mode in Fig. 2.8. The character of the motion associated with the amide I band was demonstrated by calculating the contributions of the stretching and bending of individual bonds to the total potential energy. On this basis, for the amide I vibration 80% of the potential energy is associated with the C=O stretching, 10% with C–N stretching, and 10% with an in-plane bending motion of the N–H bond. By introducing coupling factors Miyazawa (1960) and Miyazawa and Blout (1961) were then able to consider the normal modes of a polypeptide assembly of a given structure. In

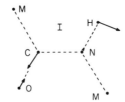

Fig. 2.8. The normal mode designated the amide I mode of the peptide linkage. The nuclei move in the directions of the arrows, and the length of the arrows is proportional to the amplitude of oscillation.

this way it was possible to demonstrate theoretically that the exact position of the amide I band differed for structures such as the polypeptide α-helical and pleated-sheet formations. Correlations of this kind have been further substantiated by experiment and the position of the amide I band in the infrared and Raman spectra of a protein can be used to give an indication of the secondary structure of the peptide backbone. In the Raman spectra, another peptide mode, designated the amide III vibration is even more useful in this respect; this topic is discussed in Chapter 4, Sections III,A and III,B.

It should be kept in mind that the group frequency approach is always an approximation and sometimes is not valid at all. The idea is often expressed in the literature that, within a certain molecule, the group frequency concept breaks down as a result of vibrational coupling. This simply means that within that molecule the group does not act as an independent oscillator in the normal mode most like the "group mode" and that nuclei not associated with the group have large amplitudes of oscillation. The linear system shown in Fig. 2.9 provides a simple model for understanding vibrational coupling. If the masses m_1, m_2, and m_3 are approximately the same and if the force constants of the springs are similar, an oscillation of m_1 and m_2 about their equilibrium positions will be strongly affected by the presence of the spring connecting m_2 and m_3 and by m_3 itself. Hence neither the nuclei m_1 and m_2 together nor m_2 and m_3 together can be regarded as independent oscillators, and they are said to be mechanically or vibrationally coupled. However, if there are sizable disparities in the force constants or masses the vibrational coupling may

Fig. 2.9. Three particles connected by weightless springs having force constants of k and k'. The system is constrained such that it remains linear and does not rotate.

become minimal. In one extreme if the mass of m_2 becomes infinitely large, then m_1 and m_3 will oscillate independently and appear as two separate mass and spring systems attached to a wall. The O–H, C–H, N–H, and S–H groups cited previously are examples of large disparities in nuclear mass leading to essentially independent oscillators. The model used in Fig. 2.9 constrains the masses to remain in a straight line; consequently, coupling occurs only between stretching vibrations. For molecules, in addition to coupling between stretching modes, coupling is frequently encountered between, for example, a bending vibration and a stretching vibration. However, molecular symmetry does impose limitations on which vibrations may couple and coupling can only occur between vibrations of the same symmetry species. For example, in a molecule possessing a plane of symmetry, such as $CH_2{=}CHBr$, in-plane vibrations do not interact with out-of-plane vibrations.

D. Isotopic Substitution

Isotopic substitution can be an invaluable aid in the analysis of vibrational spectra. By making the assumption, which is generally valid, that the force constants are unaltered by isotopic substitution, the shift in observed frequency can be attributed principally to mass effects. Then, in a given spectrum, by observing which bands shift and the extent to which they shift on isotopic replacement it is often possible to gain insight into the participation of the replacement nucleus in the normal mode giving rise to the bands. Vibrations which involve large amplitudes of the atom of the substituted species should shift appreciably, and the larger the isotopic mass difference, the greater the shifts will be. In this regard deuterium substitution gives a particularly large shift. An example is the shift in the C–H stretching frequency on forming the C–D moiety. This substitution is utilized in membrane studies to shift fatty acid lipid C–H modes to a spectral region uncluttered by modes from other membrane components (Chapter 8, Section II,A,4). Equation (2.24) can be used to calculate the expected frequency shift for a diatomic oscillator, and we can attempt to treat the C–H group as an isolated oscillator and calculate the shift on deuterium substitution. Recall Eq. (2.24)

$$\nu = \frac{1}{2\pi}\sqrt{\frac{k}{\mu}}$$

Only the reduced mass μ will change with isotope replacement

$$\therefore \nu_{C-H} = (1/2\pi)\sqrt{k/\mu_{C-H}}$$
$$\nu_{C-D} = (1/2\pi)\sqrt{k/\mu_{C-D}}$$

and

$$\nu_{C-H}/\nu_{C-D} = \sqrt{\mu_{C-D}/\mu_{C-H}}$$

Since $\mu = m_1 m_2/(m_1 + m_2)$ and $m_H = 1$, $m_D = 2$, and $m_C = 12$,

$$\nu_{C-H}/\nu_{C-D} = \sqrt{(12 \times 2/14) \times (13/12 \times 1)}$$

$$= 1.3628$$

Since ν_{C-H} for alkane chains occurs near 2900 cm^{-1}, we predict that ν_{C-D} will be found near 2100 cm^{-1}. This is in fact where the C–D stretch occurs. Usually, the treatment of a pair of neighboring atoms in a molecule as a diatomic oscillator is a gross oversimplification, but it works well in the case of the C–H stretch because, for the reasons outlined previously, modes involving the X–H nuclei are good group frequencies.

A good guide to the pre-1970 literature in the vibrational spectra of isotopically labeled compounds is the volume by Pinchas and Laulicht (1971). This work also introduces some of the more sophisticated product rules that can be used to aid interpretation.

E. Overtone, Combination, and Fermi Resonance Bands

These bands appear in infrared and Raman spectra as a result of the breakdown of the harmonic oscillator approximation. As we saw in Section VI,B, the assumption that the potential energy term U assumed the form $U \propto x^2$ (where x is a measure of the displacement from an equilibrium distance) led to the finding that the vibrational motions of a complex molecule are composed of a number of simpler components called the normal modes in which each atom performs simple harmonic motion about its equilibrium position with characteristic frequency ν_{vib}. Within the simple harmonic oscillator approximation, infrared and Raman transitions are only allowed to change the vibrational quantum number by ± 1, as shown in Fig. 2.4. For actual molecules the harmonic approximation arising from writing the potential energy term $U \propto x^2$ holds quite well. However, it is necessary to introduce a degree of *anharmonicity* to fully describe the vibrational motions. A more realistic description of the potential energy function is given by the *Morse equation* for a diatomic oscillator

$$U = D_e\{1 - \exp[-\beta(r - r_e)^2]\} \qquad (2.47)$$

where U is the potential energy as a function of the displacement from the equilibrium internuclear distance $r - r_e$, β is a constant, and D_e is the dissociation energy required to reach total separation of the nuclei starting

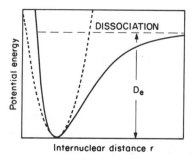

Fig. 2.10. The potential energy function of a diatomic molecule as a function of internuclear distance r. The equilibrium internuclear distance r_e is very close to the minimum in the potential curve. The broken curve is the potential energy function for a simple harmonic oscillator. Adapted from Colthup *et al.* (1975).

at the curve minimum shown in Fig. 2.10. It is easily shown that for small values of $r - r_e$

$$U = D_e \beta^2 (r - r_e) \quad \text{or} \quad U \propto (r - r_e)^2 \tag{2.48}$$

which, of course, is analogous to the potential of a harmonic oscillator $U \propto x^2$. The equivalence of the Morse and harmonic oscillator functions near the potential energy minimum is shown in Fig. 2.10. Moreover, the Morse curve fits our intuitive ideas of how the potential energy of a molecule should vary with r. At small distances the energy increases rapidly because of the close approach of the nuclei, and at large values of r, after dissociation, the energy reaches a plateau and becomes insensitive to changes in r.

If we no longer make the approximation in Eq. (2.47) that $r - r_e$ is very small, we must consider terms in the expansion containing $(r - r_e)^n$ in which n is greater than 2. These higher-order terms introduce anharmonicity into the description of the nuclei motions. As a consequence, the simple selection rule stated in Section V that vibrational transitions are only allowed between levels differing in quantum numbers by ± 1 no longer applies universally. In addition to the *fundamental* transitions with $\Delta v = \pm 1$, *overtones* can occur in which $\Delta v > \pm 1$. Turning to Fig. 2.4 we see that the vibrational transition from $v'' = 0$ to $v'' = 1$ in the ground electronic state has been demonstrated. As a result of anharmonicity a further, but generally less probable, transition from $v'' = 0$ to $v'' = 2$ can occur. Consider the resonance Raman spectrum of the molecule β-carotene shown in Fig. 2.11; for the normal mode associated with the stretching of the $C=C$ bonds an intense feature occurs at frequency $\nu_1 = 1523 \text{ cm}^{-1}$ due to the transition from the $v'' = 0$ to the $v'' = 1$ vibrational

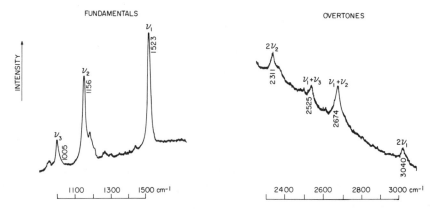

Fig. 2.11. The resonance Raman spectrum of β-carotene showing the fundamental and overtone regions. The β-carotene is dissolved in CCl_4 and the overtone region is recorded at 2 times the gain used for the fundamentals.

level in the ground electronic state. However, a less intense *overtone* feature also appears at approximately $2\nu_1$ due to the transition from $v'' = 0$ to $v'' = 2$. The exact position of this overtone is at 3040 cm^{-1} since the anharmonicity, in addition to making the overtone an "allowed" transition, perturbs the spacing of the vibrational energy levels slightly; hence the spacings in Fig. 2.4 actually narrow as the v quantum number increases. An overtone attributable to the ν_2 mode occurs at 2311 cm^{-1}.

Anharmonicity causes the appearance not only of overtones, but also of weak *combination* bands as well. A combination band appears near the frequency of the sum of two (or occasionally three) fundamental bands. For example, the features at 2674 and 2525 cm^{-1} in the resonance Raman spectrum of β-carotene shown in Fig. 2.11 arise from $\nu_1 + \nu_2$ and $\nu_1 + \nu_3$, respectively.

Within the harmonic oscillator approximation each normal mode of a polyatomic molecule has at most one vibrationally active transition from the $v'' = 0$ ground vibrational state. As seen previously, the introduction of anharmonicity may give rise to additional overtone or combination bands. Although these extra bands are usually weak, it sometimes happens that the frequency of an overtone or combination band nearly coincides with the frequency of a fundamental band. In this case two relatively intense bands may be observed where only one strong band for the fundamental was expected. These are observed at somewhat higher and lower frequencies than the predicted band positions of the fundamental and overtone. This effect is known as *Fermi resonance*. Both the observed bands contain contributions from the fundamental and the overtone.

Anharmonicity leads to the mixing of the fundamental and overtone transitions. Overtone, combination, and Fermi resonance bands are observed in infrared and Raman spectra, and Fermi resonance is in no way connected with the resonance Raman effect.

An example of the occurrence of a Fermi doublet which has diagnostic value in biochemistry is the pair of lines near 830 and 850 cm^{-1} observed in the Raman spectra of many proteins. This doublet is caused by the phenol side chain of the tyrosine amino acids in the protein, and the utility of the doublet stems from the fact that the relative intensity of the two peaks is sensitive to environment. A good model for the tyrosine side chain is p-cresol, and Fig. 2.12 shows the occurrence of the doublet for p-cresol in H$_2$O solution. The sensitivity of the doublet to change in state of the phenolic moiety is demonstrated by the intensity reversal on ionization of the —OH group. A collaborative paper by groups at the Massachu-

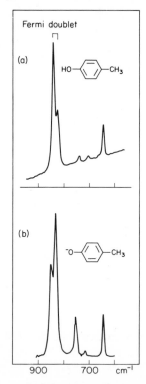

Fig. 2.12. Raman spectrum of p-cresol showing that the relative intensity of the Fermi doublet near 840 cm^{-1} depends on the state of ionization of the —OH group. (a) Aqueous solution pH 6.7. (b) Aqueous solution pH 13. Reprinted with permission from M. N. Siamwiza *et al.*, *Biochemistry* **14**, 4870 (1975). Copyright 1975 American Chemical Society.

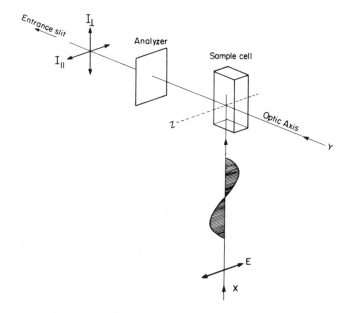

Fig. 2.13. The orientations of the **E** vectors of the incoming (**E**) and scattered light parallel ($I_\parallel$) and perpendicular ($I_\perp$) to the incident **E** vector. The depolarization ratio $\rho = I_\perp/I_\parallel$.

setts Institute of Technology in the United States and the University of Tokyo in Japan (Siamwiza *et al.,* 1975) demonstrated that the Fermi doublet arises from interaction of the overtone from a nonplanar ring vibration at 413 cm^{-1} with a symmetric ring-breathing fundamental near 840 cm^{-1}.

VII. Polarization Properties of Raman Scattering

According to the electromagnetic theory, any type of light consists of transverse waves, in which the oscillating magnitudes are the electric and magnetic vectors. Most natural light is unpolarized, which in a simplified form means the electric (**E**) vector performs linear oscillations of constant amplitude in a plane perpendicular to the light path but that the orientation of the **E** vector within that plane is completely random. Light from lasers has the special property of being plane polarized; that is, the terminus of the electric vector varies periodically within a single plane through the light path. Thus, for the plane polarized beam traveling along the x axis in Fig. 2.13, the **E** vector remains in the xz plane, but varies in magnitude along the light path according to the hatched lines. Information on the scattering process and on the assignments of Raman bands can be gained by analyzing the scattered light parallel and perpendicular to the incoming **E** vector.

The *depolarization ratio* ρ of a feature in the Raman spectrum is defined as

$$\rho = I_\perp/I_\parallel \qquad (2.49)$$

where $I_\perp$ and $I_\parallel$ are the intensities of Raman radiation of a given frequency that is polarized, respectively, perpendicular and parallel to the plane normal to the incident beam. These relationships are shown in Fig. 2.13. Changes in the polarization of the incident light on scattering result from the tensorial nature of the interaction denoted in Eq. (2.34). In fact, for a single crystal the individual elements of the scattering tensor α may be related to $I_\perp$ and $I_\parallel$.* However, for fluids the molecules are randomly oriented with respect to the laboratory-fixed coordinate system used to define α, and when an average taken over molecular orientations is made, $I_\perp$ and $I_\parallel$ are found to be related to certain combinations of the components of α. These combinations are known as the tensor invariants. Thus, for randomly oriented molecules, when the intensities $I_\perp$ and $I_\parallel$ are computed, they are found to be

$$I_\perp = \text{const} \times (3g^s + 5g^a)\bar{E}_z^2 \qquad (2.50)$$

and

$$I_\parallel = \text{const} \times (10g^0 + 4g^s)\bar{E}_z^2 \qquad (2.51)$$

where, as in Fig. 2.13, the incident light is taken to be polarized in the z direction; $\bar{E}_z^2$ is the average value of the square of the electric vector, to which the intensity of the incident light is proportional. The depolarization ratio is therefore given by

$$\rho = (3g^s + 5g^a)/(10g^0 + 4g^s) \qquad (2.52)$$

where the tensor invariants are

$$g^0 = \tfrac{1}{3}(\alpha_{xx} + \alpha_{yy} + \alpha_{zz})^2$$

$$g^s = \tfrac{1}{2}\{(\alpha_{xx} - \alpha_{yy})^2 + (\alpha_{xx} - \alpha_{zz})^2 + (\alpha_{yy} - \alpha_{zz})^2\}$$

$$\qquad + \tfrac{3}{4}\{(\alpha_{xy} + \alpha_{yx})^2 + (\alpha_{xz} + \alpha_{zx})^2 + (\alpha_{yz} + \alpha_{zy})^2\}$$

$$g^a = \tfrac{3}{4}\{(\alpha_{xy} - \alpha_{yx})^2 + (\alpha_{xz} - \alpha_{zx})^2 + (\alpha_{yz} - \alpha_{zy})^2\} \qquad (2.53)$$

---

* Depolarization measurements can be made for single crystals, but not for polycrystalline solids since the many changes in refractive index experienced by the light beam traversing polycrystalline materials tend to "scramble" the **E** vector.

The entities g^0, g^s, and g^a are known as the isotropic, symmetric aniso-tropic, and antisymmetric anisotropic components, respectively. Always under nonresonance conditions, and usually for resonance conditions, $g^a = 0$. This being the case

$$\rho = 3g^s/(10g^0 + 4g^s) \qquad (2.54)$$

and since in vibrational Raman scattering either g^s or g^0 can be zero, it follows from Eq. (2.54) that $0 \leq \rho \leq 0.75$. Detailed calculations show that for normal vibrations which do not preserve molecular symmetry during the motion of the nuclei (non-totally-symmetric modes) $g^0 = 0$, and hence from Eq. (2.54) $\rho = 0.75$. For modes which do preserve molecular sym-metry neither g^0 nor g^s is zero leading to values of $\rho \leq 0.75$ for these modes. In fact, ρ is often found to be substantially less than 0.75 for totally symmetric vibrations.

> For totally symmetric modes $\rho \leq 0.75$
>
> For non-totally-symmetric modes $\rho = 0.75$

Carbon tetrachloride provides a good example of the relation between symmetric and non-totally-symmetric modes and depolarization ratios. The totally symmetric mode ν_1 shown in Fig. 2.14 preserves the symmetry of the molecule; consequently, $\rho < 0.75$. Actually, this feature, occurring

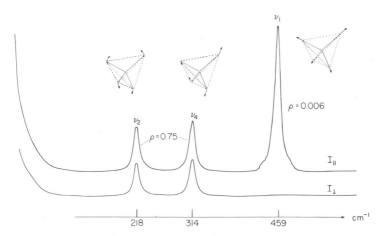

Fig. 2.14. The Raman spectra of CCl_4 recorded with the analyzer in the $I_\parallel$ and $I_\perp$ orienta-tions (see Fig. 2.13). The intensity of ν_1 in the $I_\perp$ spectrum is too low to be observed in the recording shown. The normal modes giving rise to ν_2, ν_4, and ν_1 are shown beside the respective Raman peaks.

at 459 cm^{-1}, is highly polarized with $\rho = 0.006$. In contrast, the motion of the nuclei in the ν_2 and ν_4 modes of frequency 218 and 314 cm^{-1} (Fig. 2.14) do not preserve the tetrahedral symmetry of CCl_4, and $\rho = 0.75$ for both these non-totally-symmetric vibrations. As is mentioned in Chapter 3, Section I,B,3, these vibrations provide a good test for checking that the Raman instrument is correctly set up for measuring depolarization ratios.

Under resonance or near resonance conditions in certain rather rare instances $g^a \neq 0$. This arises because for certain modes α can become antisymmetric with $\alpha_{\sigma\rho} = -\alpha_{\rho\sigma}$, giving rise to "anomalously" polarized bands with $\rho > 0.75$. In the limiting case when $g^s = g^0 = 0$, but $g^a \neq 0$, the phenomenon of inverse polarization occurs with, in theory, $\rho = \infty$. The discovery, by Spiro and Strekas (1972), of anomalously polarized lines in the resonance Raman spectra of some heme proteins awakened many spectroscopists interest in biological molecules since this was the first report of the effect in vibrational Raman spectroscopy. Interestingly, Placzek (1934), in his classic monograph, had predicted the occurrence of anomalous polarization some 40 years prior to the experimental finding. In the usual case when $g^a = 0$, the relative values of the *two* tensor invariants, g^0 and g^s can be determined from Eq. (2.54) by the measurement of ρ. However, for $g^a \neq 0$ there are obviously *three* tensor invariants which, in general, cannot be determined by measurement of ρ alone. To determine these three invariants, it is necessary to use circularly polarized incident light and to analyze the polarizations of the forward or backward scattered radiation, (Nestor and Spiro, 1973; Pézolet *et al.*, 1973).

VIII. Raman Intensities

By deriving Eq. (2.42) using classical theory we were able to show that the interaction of a light wave of frequency ν_0 with a diatomic molecule m_1m_2 vibrating with frequency ν_{vib} can give rise to inelastically scattered light of frequencies $\nu_0 - \nu_{vib}$ and $\nu_0 + \nu_{vib}$. However, the classical equation (2.42) only supplies information regarding the energy of the scattered light. To evaluate its intensity I, we have to use the relation

$$I = K(\nu_0 - \nu_{vib})^4 I_0 (d\alpha_{zz}/dr)^2 \qquad (2.55)$$

Here we are considering the Stokes process involving the tensor component α_{zz} and where K is a constant, I_0 is the incident light intensity and r the distance separating the m_1 and m_2 nuclei. Equation (2.55) demonstrates that the intensity of Raman light is proportional to the fourth power of the *scattered* light. This gives rise to the so-called ν^4 law and demonstrates that there is a natural barrier to extending Raman studies toward the infrared region of the optical spectrum since the ν^4 factor becomes small with diminishing values of ν. Additionally, it is apparent that the intensity depends on the change in polarizability with the change in r, caused by the molecule vibrating along the bond linking m_1 to m_2. This is the basis of bond polarizability theory (Hester, 1967), which seeks to relate Raman intensities to the polarizabilities of a chemical bond, or bonds, associated with a normal mode. Although this model, based on the classical approach, is only concerned with the properties of the ground electronic state, it provides useful insight into intensities under conditions far from resonance. Some rules of thumb emerging from these consid- erations and relating intensities to chemical properties have been sum- marised by Nishimura *et al.* (1978), and they are listed at the end of this section. In order to gain further insight into resonance enhancement, how- ever, we must turn to a quantum mechanical approach in which excited electronic states become explicitly involved. The next few pages contain some equations which the biochemists may find alarming. The brave can read on and concentrate on the discussion stemming from the mathemat- ics. The less brave can turn to the end of the section, where the conclu- sions are summarized.

In the quantum mechanical treatment the factor $d\alpha_{zz}/dr$ [or in more general terms $\partial\alpha_{\sigma\rho}/\partial Q$, where $\alpha_{\sigma\rho}$ is one of the nine elements of the tensor in Eq. (2.34) and Q is a normal coordinate] is replaced by the transition polarizability tensor $|\alpha_{\sigma\rho}|_{mn}$, which may be derived by second-order per- turbation theory. Thus, for randomly oriented molecules, the total inten- sity of the scattered light resulting from a molecular transition between states m and n is

$$I = K(\nu_0 \pm \nu_{mn})^4 I_0 \sum_{\sigma\rho} |(\alpha_{\sigma\rho})_{mn}|^2 \qquad (2.56)$$

where

$$(\alpha_{\sigma\rho})_{nm} = \frac{1}{h} \sum_r \frac{\langle n|\mu_\rho|r\rangle\langle r|\mu_\sigma|m\rangle}{\nu_{rm} - \nu_0 + i\Gamma_r} + \frac{\langle n|\mu_\sigma|r\rangle\langle r|\mu_\rho|m\rangle}{\nu_{rn} + \nu_0 + i\Gamma_r} \qquad (2.57)$$

In these expressions the molecule is considered to be in the molecular state m. It is perturbed by an electromagnetic wave of frequency ν_0 and intensity I_0, causing the transition to a state n and scattering light of frequency $\nu_0 \pm \nu_{mn}$. The sum over index r covers all of the quantum mechanical eigenstates of the molecule, h is Planck's constant, and Γ_r is a damping constant which takes into account the finite lifetime of each molecular state. The $\langle n|\mu_\rho|r\rangle$, etc., are the amplitudes of the electric dipole transition moments, where μ_ρ is the electric dipole moment operator along direction ρ. Immediately, we see from Eq. (2.57) that as ν_0

approaches the energy of an allowed molecular transition ν_{rm}, ($\nu_{rm} - \nu_0 + i\Gamma_r$) becomes small; consequently, one term in the sum becomes very large. This is the resonance condition. To examine this more closely, the molecular states m and n have to be broken down into their electronic and vibrational components. This is achieved within the Born–Oppenheimer approximation by writing the total wave functions as products of electronic and vibrational parts. Thus

$$|m\rangle = |g\rangle|i\rangle, \qquad |n\rangle = |g\rangle|j\rangle, \qquad |r\rangle = |e\rangle|v\rangle \qquad (2.58)$$

where g is the ground and e the excited electronic state, i, j are vibrational states associated with g, and v is a vibrational state associated with e. We can see an exact analogy with the energy levels depicted in Fig. 2.4. The lower electronic state corresponds to g and the upper electronic state to e; i and j are v'' levels in g, and v corresponds to a v' level in e. Using Eq. (2.58) the dipole transition moments now become

$$\langle n|\mu_\rho|r\rangle = \langle j|M_e^\rho|v\rangle \quad \text{and} \quad \langle r|\mu_\sigma|m\rangle = \langle v|M_e^\sigma|i\rangle \qquad (2.59)$$

The expressions on the right of the equal signs in Eq. (2.59) now involve only *vibrational* wave functions. The electronic wave functions appear in M_e, which is the pure electronic transition moment from the ground to the eth excited electronic state. M_e changes as the nuclear coordinates vary, and this may be taken into account by expanding the electronic transition dipole moment in a Taylor series about the equilibrium geometry as a function of the normal coordinate Q.

$$M_e = M_e^0 + (\partial M_e/\partial Q)^0 Q + \cdots \qquad (2.60)$$

There are equivalent expressions for each of the $3N - 6$ possible normal coordinates. After introducing the expression for M_e from Eq. (2.60) into Eq. (2.59) and substituting into Eq. (2.57) we have

$$(\alpha_{\sigma\rho})_{mn} = A + B \qquad (2.61)$$

where

$$A = \frac{M_e^\rho M_e^\sigma}{h} \sum_v \frac{\langle j|v\rangle\langle v|i\rangle}{\nu_{vi} - \nu_0 + i\Gamma_v} \qquad (2.62)$$

and

$$B = \frac{M_e^\rho(\partial M_e^\sigma/\partial Q)^0}{h} \sum_v \frac{\langle j|v\rangle\langle v|Q|i\rangle}{\nu_{vi} - \nu_0 + i\Gamma_v} \qquad (2.63)$$

The ν_{vi} term is the energy gap between the ith vibrational level in the ground state and vth vibrational level in the excited resonant electronic state. The nonresonant term in Eq. (2.61) has been ignored, and a second term on the right-hand side of the equation for B, in which ρ and σ are reversed, has been omitted.

Usually, the A term is the dominant contributor to resonance Raman intensity since for allowed transitions M_e, the electronic transition moment, is larger than $\partial M_e / \partial Q$. Only totally symmetric modes give rise to resonance enhancement via the A term because for non-totally-symmetric modes the vibrational overlap integrals $\langle j|v \rangle \langle v|i \rangle$ (also known as the *Franck–Condon factors*) are equal to zero. The number of modes showing enhancement through the A term is further reduced by the fact that high band intensities are only observed from those modes in which the variation of nuclei positions correspond to the distortions experienced by the molecule when going to its resonant excited state (Hirakawa and Tsuboi, 1975). For example, as can be seen in Fig. 2.11, in the case of carotenoid molecules the $\pi \rightarrow \pi^*$ transition leads to large intensity gain for those modes at 1523 and 1156 cm^{-1} involving stretching of the C=C and C—C bonds, respectively, since these bond distances change appreciably in the π^* state. However, the C—H bonds are little perturbed on going from the π to the π^* state; consequently, C–H stretching modes show minimal intensity variation as the resonance condition is approached. In fact under the conditions used to record Fig. 2.11 the C–H modes near 3000 cm^{-1} cannot be observed above the spectral background.

Unlike the A term, which involves a single excited electronic state, the B term arises from the vibronic mixing of two excited states. Although the active vibration may have any symmetry which is contained in the direct product of the group-theory representations of the two electronic states (Johnson and Peticolas, 1976), the major importance of the B-term mechanism is the intensity it allows to scattering by non-totally-symmetric modes. Vibrations will be observed that are effective in "borrowing" intensity for a weak transition from a nearby intense transition. For example, excitation into a $n \rightarrow \pi^*$ band that is vibronically coupled to an intense $\pi \rightarrow \pi^*$ transition leads to enhancement of modes via the B-term mechanism. An *excitation profile* is the dependence of the intensity of a Raman band on excitation wavelength, and such a profile can be useful in assigning features in an absorption spectrum. For example, an interesting consequence of the B-term mechanism is that excitation profile maxima are expected at 0–0 and 0–1 positions for each vibronically active mode since the products $\langle j|v \rangle \langle v|Q|i \rangle$ and $\langle j|Q|v \rangle \langle v|i \rangle$ (Eq. 2.63) are largest for these levels (Spiro and Stein, 1977; Strekas and Spiro, 1973). Examples of excitation profiles and their relation to absorption spectra are given in Chapter 5, Sections II and V, and Chapter 7, Section V.

By applying several simplifying approximations Albrecht and Hutley (1971) have been able to show that the A and B terms show quite different frequency dependencies in the preresonance region. For the A-type mechanism

$$I \propto (\nu_0 \pm \nu_{ij})^4 \left[\frac{\nu_e^2 + \nu_0^2}{(\nu_e^2 - \nu_0^2)^2} \right]^2 \equiv F_A^2 \tag{2.64}$$

while for B terms

$$I \propto 4 \left(\nu_0 \pm \nu_{ij}\right)^4 \left[\frac{\nu_e \nu_s + \nu_0^2}{(\nu_e^2 - \nu_0^2)(\nu_s^2 - \nu_0^2)} \right]^2 \equiv F_B^2 \qquad (2.65)$$

where ν_{ij} is the Raman frequency shift, I is the intensity of the scattered light for the vibrational transition for i to j, ν_0 the excitation frequency, ν_e is the frequency of the electronic transition to the resonant excited state and ν_s is the frequency of electronic transition to the second excited state utilized in the B-type mechanism. The theoretical simplifications needed to derive Eqs. (2.64) and (2.65) are drastic and moreover, the preresonance region, in which the equations are valid, is not always clearly defined. However, relationships of this kind do provide empirical criteria to compare the observed frequency dependencies of band intensities.

Some Simplified Generalizations Concerning Raman Intensities

From the foregoing it is apparent that the consideration of the intensities of features in a Raman spectrum involves parameters, such as the description of the excited electronic state, that are hard to quantitate. Indeed, the topic of resonance Raman enhancement is an active area for theoreticians (Clark and Stewart, 1979; Johnson and Peticolas, 1976; Siebrand and Zgierski, 1979; Spiro and Stein, 1977; Warshel, 1977). It is still possible, however, to draw some simple generalizations together to aid in a qualitative understanding of intensities. These should not be regarded as universal, unequivocal laws, but rather as useful rules of thumb and, for this reason, should be used with care.

The classical approach to Raman intensities summarized by Eq. (2.55) draws attention, in the normal Raman case, i.e., far from resonance, to the importance of polarizability of the electrons associated with chemical bonds or groupings. The Raman intensity depends on the degree of change of the polarizability caused by a change in normal coordinate. On this basis Nishimura *et al.* (1978) suggested the following:

(a) Stretching vibrations associated with chemical bonds should be more intense than deformation vibrations.

(b) Multiple chemical bonds should give rise to intense stretching modes, e.g., a Raman band due to a $C{=}C$ vibration should be more intense than that due to a $C{-}C$ vibration.

(c) Bonds involving atoms of large atomic mass are expected to give rise to stretching vibrations of high Raman intensity. The S–S linkages in proteins are good examples of this.

(d) Those Raman features arising from normal coordinates involving two in-phase bond stretching motions are more intense than those involving a 180° phase difference. Similarly, for cyclic compounds the in-phase "breathing" mode is usually the most intense. The intense phenyl ring mode of phenylalanine which occurs near 1005 cm^{-1} in the Raman spectra of proteins is a good example of an intense in-phase breathing mode.

Under resonance conditions startling Raman intensity enhancement of certain modes may be observed with increases of the order of 10^4 over intensities obtained far from resonance. There are two main scattering mechanisms giving rise to this enhancement, one of which, the A-term mechanism, involves a single excited electronic state whereas the second, or B-term mechanism, involves two excited electronic states. The two mechanisms may be distinguished on the basis of their different frequency dependencies [Eqs. (2.64) and (2.65)]. Large intensity enhancements are expected when

(e) The absorption intensity of the electronic transition is large and when the absorption peak has a narrow profile.

(f) The excitation frequency is close to that of the electronic transition.

Considerations e and f hold for both A-term and B-term mechanisms.

(g) For the A-term mechanism only, some of the products of the Franck–Condon factors $\langle j|v\rangle\langle v|i\rangle$ are large.

(h) For the B-term mechanism only, the vibration is effective in mixing the two excited electronic states.

(i) The geometric changes experienced during a particular normal mode are similar to those changes occurring when going from the ground to the excited electronic states. This rule holds for the A-term mechanism and, according to Nishimura *et al.* (1978), for the B-term mechanism also.

It is important to emphasize that the positions of peaks in Raman and resonance Raman spectra are a property solely of the electronic ground state. While Raman peak intensities may be correlated with ground state properties, intensities under resonance conditions can only be understood by considering the excited electronic state of the molecule. By the same token, a study of resonance Raman intensities can provide unique information regarding excited states, and, in the biological context, this holds great promise for increasing the understanding of processes, such as photosynthesis and vision, which use excited states to perform chemical transformations.

CHAPTER 3

Experimental Raman Spectroscopy

I. Basic Optics of the Raman Experiment

In a standard Raman experiment, intense monochromatic radiation provided by a laser is focused onto or into the sample. Some of the resulting scattered light is gathered by collecting optics and directed to a dispersing system, usually a double monochromator. The function of the monochromator is to separate spatially the scattered light on the basis of frequency. At the exit port of the monochromator the Raman spectrum forms an image in the form of a series of very faint lines. These are detected and recorded either sequentially by a single photomultiplier used with a scanning monochromator or simultaneously by a multichannel detector, which is the modern electronic equivalent of a photographic plate. These methods of detection are discussed further below.

A schematic of the important prespectrometer optics in a Raman experiment is shown in Fig. 3.1. Starting at the laser, we shall work our way along the light path and consider the key facets of each component in turn.

A. The Laser

A laser (light amplification by stimulated emission of radiation) is a device that depends on emission by stimulated processes. The argon gas laser is used as an example. Argon gas is contained in an evacuated glass

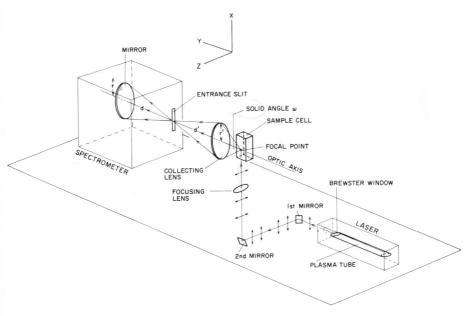

Fig. 3.1. The optics about the sample for a conventional Raman experiment.

tube, known as the plasma tube, which is held between two mirrors. Some of the argon atoms are excited by an electrical discharge to become ionized Ar^+ species which spontaneously emit a photon. The emitted photon ricochets backward and forward along the tube and stimulates other argon atoms to emit photons. They add more photons of the same frequency to the cavity formed by the tube and the mirrors, and these photons stimulate more atoms to emit photons. The stimulated emission builds up rapidly, and by making one of the mirrors partially transparent the light source may be tapped. The characteristics of laser radiation follow from the means of its generation; it is monochromatic (because photons stimulate photons of the same frequency), unidirectional (photons not traveling along the axis of the cavity are lost), and coherent (the physics of the stimulation process requires that the phases of the electric field of the photons be in step).

Most Raman experiments use continuous wave (CW) lasers which emit a continuous stream of light. Among the continuous wave sources, argon and krypton ion gas lasers are by far the most widely used systems. Argon and krypton lasers can be tuned between various discrete wavelengths, and these available laser lines along with typical power outputs are given in Table 3.1. Alternatively, continuous tuning within a given

Table 3.1

Continuous Wave Lasers for Raman Excitation[a]

Wavelength in air (nm)	Wavenumber in air (cm^{-1})	Wavenumber in vacuum (cm^{-1})	Relative output powers (mW)			
			Kr	Ar	He–Ne	He–Cd
799.32	12510.6	12507.2	30[b]	—	—	—
793.14	12608.1	12604.6	10[b]	—	—	—
752.55	13288.2	13284.5	150[b]	—	—	—
676.44	14783.2	14779.2	120	—	—	—
647.09	15453.9	15449.6	750	—	—	—
632.82	15802.3	15798.0	—	—	100	—
568.19	17599.8	17594.9	200	—	—	—
530.87	18837.2	18831.9	250	—	—	—
528.69	18914.7	18909.4	—	200[b]	—	—
520.83	19200.1	19194.7	150	—	—	—
514.53	19435.1	19429.7	—	800	—	—
501.72	19931.6	19926.0	—	140	—	—
496.51	20140.7	20135.1	—	300	—	—
487.99	20492.4	20486.7	—	700	—	—
482.52	20724.7	20718.9	75	—	—	—
476.49	20987.0	20981.1	—	300	—	—
476.24	20997.7	20991.8	50	—	—	—
472.69	21155.7	21149.8	—	60	—	—
468.04	21365.7	21359.8	500[c]	—	—	—
465.79	21468.9	21462.9	—	50	—	—
457.94	21837.2	21831.1	—	150	—	—
454.51	22002.0	21995.8	—	20	—	—
441.56	22646.8	22640.5	—	—	—	50
415.44	24070.9	24064.2	100[c]	—	—	—
413.13	24205.5	24198.8	1,500[c]	—	—	—
406.74	24585.7	24578.8	900[c]	—	—	—
379.53	26348.2	26340.8	—	20[c]	—	—
363.79	27488.5	27480.6	—	700[c]	—	—
356.42	28056.6	28048.5	270[c]	—	—	—
351.42	28456.1	28448.0	—	10[c]	—	—
351.11	28480.9	28472.8	—	600[c]	—	—
350.74	28511.0	28502.8	1,300[c]	—	—	—
337.50	29629.6	29621.2	170[c]	—	—	—
335.85	29775.3	29766.7	—	10[c]	—	—
334.47	29897.9	29889.3	—	30[c]	—	—
333.61	29974.9	29966.2	—	80[c]	—	—
325.03	30766.4	30757.7	—	—	—	15[b]
323.95	30869.0	30860.3	25[c]	—	—	—

[a] Powers quoted are for lasers of nominal total output of about 2 W.

[b] Requires special mirror optics.

[c] Requires high-power laser and special mirrors; however, some UV lines can be obtained from 3–5-W lasers. In all cases the UV lines have to be separated outside the cavity.

spectral range can be achieved by using a dye laser. In essence, a dye laser consists of a light source (e.g., an argon or krypton laser) which "pumps" the fluorescence spectrum of a dye. A narrow spectral bandwidth within this fluorescence output is made to lase in the dye laser cavity, and since the total fluorescence output covers a broad spectral bandwidth (typically 50 nm), laser output can be tuned continuously within the overall fluorescence emission.

Some laser sources suitable for Raman experiments emit a train of short-lived high-power pulses. One advantage of pulsed lasers is that they can be frequency doubled by passing the beam through certain inorganic crystals. Thus pulses of light of 600-nm wavelength can be converted, with an efficiency of a few percent, to 300-nm wavelength by passage through an ammonium dihydrogen phosphate crystal. Doubling efficiency is generally proportional to the square of the power; thus frequency doubling is only a tractable proposition for pulsed lasers since these have very high power within the time span of each pulse. In experiments using pulsed lasers the Raman photons arrive at the photomultiplier tube in pulses, and the resulting signal is normally analyzed using a box car integrator.

1. Light energy and photon flux

It is sometimes necessary to know the energy of a photon or a "mole" (i.e., Avogadro's number) of photons of a given wavelength. Conversely, given a certain laser power it may be useful to calculate how many photons are delivered per second or per pulse in the case of a pulsed laser source. The basic formula relating energy to wavelength is

$$E = h\nu \qquad (3.1)$$

where h is Planck's constant, 6.626×10^{-34} J sec. Thus, using $\nu = c/\lambda$ (Chapter 1, Section III) one photon at 500 nm has an energy of

$$E = 6.626 \times 10^{-34} \times \frac{2.998 \times 10^8}{500 \times 10^{-9}} \text{ J}$$
$$= 3.973 \times 10^{-19} \text{ J}$$

or generally the energy of a single photon E_p is

$$\boxed{E_p = \frac{1.986 \times 10^{-16}}{\lambda \text{ (nm)}} \text{ J}} \qquad (3.2)$$

In terms of energy units of kilocalories per mole we can calculate the photon energy available from 1 "mole" of photons as

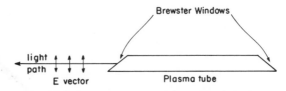

Fig. 3.2. The side-on view of a laser plasma tube showing the direction of the E vector of the emergent light beam.

$$\frac{\text{joules} \times \text{Avogadro's number}}{4.184 \times 10^3} \quad \text{(in kcal mole}^{-1}) \quad (3.3)$$

where 1 cal (thermochemical) = 4.184 J. By setting Avogadro's number equal to 6.022×10^{23} we can calculate the energy of 1 "mole" of photons at 500 nm as equal to

$$\frac{(3.973 \times 10^{-19}) \times (6.022 \times 10^{23})}{4.184 \times 10^3} = 57.18 \text{ kcal mole}^{-1}$$

The energies of 1 "mole" of photons of various wavelengths are given in Table 1.1 (Chapter 1, Section III) in terms of kJ mole^{-1} and kcal mole^{-1}.

To calculate the number of photons in a given laser pulse or delivered per second from a continuous laser, the energy in joules (remembering in the case of a continuous source that 1 watt = 1 joule per second) is divided by the energy per photon at that wavelength. Thus, by using Eq. (3.2), 500 mW of 514.5-nm excitation, e.g., from an argon ion laser, delivers

$$\frac{0.5 \times 514.5}{1.986 \times 10^{-16}} = 1.295 \times 10^{18} \text{ photons sec}^{-1}$$

2. The direction of the E vector

It is necessary to know the direction of the electric (E) vector of the light beam emerging from the laser. The E vector, represented in Figs. 3.1 and 3.2 by a double-headed arrow, emerges from the plasma tube in the plane containing the top and the bottom of the Brewster windows. For example, when the plasma tube (viewed side on) appears as in Fig. 3.2, the E vector emerges in the vertical plane. It is often necessary to guide the laser beam to the sample by using a series of mirrors, and the consequent change in direction of the E vector is shown in Fig. 3.1. The beam strikes the first mirror at 45° and remains in the horizontal plane with the E vector vertical. However, the second 45° mirror which sends the beam into the vertical causes a 90° rotation of the E vector as seen in Fig. 3.1. For all sampling geometries it is imperative that the direction of the E vector with respect to the spectrometer and sampling optics is as shown in Fig. 3.1. The E vector must not lie along the optic axis since this results in the loss of the isotropic part of the scattering process. The isotropic component of the tensor (go see Chapter 2, Section VII) induces an electric dipole in the same direction as the E vector. The direction of the light scattered from this oscillating dipole is such that none is scattered in the direction of the E vector. Hence, if the E vector is perpendicular to the direction shown in Fig. 3.1, then no Raman light attributable to the isotropic component will be scattered in the direction of the spectrometer's optic axis.

B. Optics around the Sample

1. Focusing lens

Spike filters, or a laser filter monochromator which passes the excitation wavelength but blocks weaker nonlasing lines from the laser plasma, can be inserted anywhere in the light path before the focusing lens. For strongly scattering samples, filters are used to provide a clean Raman spectrum uncluttered by laser–plasma lines—although the latter can provide excellent frequency standards for the calibration of Raman peak positions. For very accurate depolarization ratio measurements a polarizer is also placed before the focusing lens. The polarizer eliminates any light whose E vector differs in direction from that of the required orientation. For most depolarization measurements of biological molecules a polarizer is not essential.

The focusing lens increases photon density at the sample and thus increases the Raman signal. Lenses with focal lengths in the 30–70-mm range generally give a suitable focal volume in the sample. The criterion for suitability is that the image of the focal volume formed by the collecting lens should just fill the entrance slit of the spectrometer.

2. Placement of the collecting lens and the sample

The focal point of the laser in the sample and the center of the collecting lens must be on the optic axis of the spectrometer. The collecting lens should have a low f number so that it subtends a large solid angle about the focal volume and collects the optimum amount of scattered light. Moreover, the collecting lens, of radius r', should be placed at a distance d' in front of the slit such that $d'/r' = d/r$ (Fig. 3.1), where d is the distance from the slit and r the radius of the first optical element (usually a mirror) behind the slit. If this relationship holds, light from sample will just fill the spectrometer optics. Overfilling results in loss of potential Raman signal, and underfilling diminishes the resolving power of the spectrometer. The analyzer for depolarization measurements is placed between the collecting lens and the entrance slit.

3. Measuring depolarization ratios

A simple method of measuring depolarization ratios is to employ the sampling geometry shown in Fig. 2.13. For all depolarization measurements, and for most routine spectral runs, a calcite wedge is placed over the spectrometer's entrance slit. The wedge is employed to scramble the polarization of the light entering the slit and is necessary since spectrometers usually have different sensitivities toward light polarized in different directions. Using the arrangement shown in Fig. 2.13, a polarization analyzer, e.g., a Polaroid sheet, is placed between the slit and the collecting lens. The Polaroid analyzer is held on a mount that can be rotated by exactly 90°. With the analyzer oriented so that light is transmitted only in the direction of the E vector of the light striking the sample, the measured intensity is termed $I_\parallel$; by rotating the analyzer through 90° $I_\perp$ is recorded. The depolarization ratio ρ is then $I_\perp/I_\parallel$. The Raman peaks of CCl_4 make a good standard for setting up and checking the analyzer. The CCl_4 peaks at 314 or 218 cm^{-1} are depolarized and have depolarization ratios of 0.75 ± 0.02 whereas the peak at 459 cm^{-1} is strongly polarized and has a ratio of 0.006 ± 0.001 (see Fig. 2.14). The latter measurement should be made by taking the area of the peaks, but height measurements are sufficient for the depolarized bands. The Polaroid analyzer can be used to confirm that the scrambler is functioning properly by monitoring the signal from a depolarized source, e.g., a tungsten lamp. The signal should be independent of the orientation of the analyzer.

C. The Spectrometer and Detection System

The function of the spectrometer is to separate spatially the scattered photons coming from the sample on the basis of their frequency. The Raman spectrum is thus displayed as a series of faint lines across the exit port of the spectrometer. The spectrometer, shown schematically in Fig. 3.3, is a basic single monochromator. Although Fig. 3.3 shows the essential optics, in most Raman experiments the dispersing process is repeated by linking two single monochromators to form a double monochromator. A double monochromator is often necessary to separate the Raman photons from the overwhelming number of Rayleigh photons.

As shown in Fig. 3.3, there are two main methods by which the line spectrum across the exit port may be detected. The first and most commonly used method is to use a scanning spectrometer and to place a

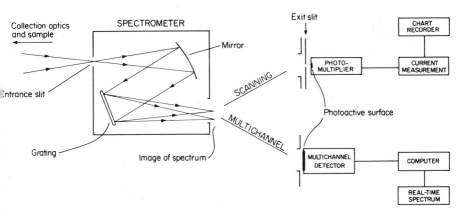

Fig. 3.3. Schematic of a Raman spectrometer showing the options of single channel (scanning) and multichannel detection.

narrow exit slit over the exit port, followed by a photomultiplier tube. By slowly turning the grating, using the accurate drive of the spectrometer, the lines of the spectrum move in succession across the slit and are detected and recorded as outlined in Fig. 3.3. In the second method a multichannel detector is placed at the exit port. This detector is similar to having several hundred minute photomultiplier tubes across the port. All the Raman lines are then registered on different elements of the detector all the time. Thus it is possible to observe the entire Raman spectrum on a television screen or an oscilloscope in real time. When using multichannel detection the grating is turned only to change the spectral region across the detector.

Scanning spectrometers

Since the majority of laboratories are equipped with scanning monochromators the following discussion focuses on these. The widths of the spectrometer slits shown in Fig. 3.3 are usually controlled manually by micrometer drives. However, the important spectroscopic quantity is not the mechanical width of the slits but the *spectral slit width*. The latter is defined as the measured bandwidth at half height in cm^{-1} of an infinitely narrow source line. The spectral slit width is a measure of the resolving power of the spectrometer and varies with the absolute wavelength at which the instrument is set and, of course, the mechanical slit widths. The *spectral slit* should always be quoted when reporting Raman spectra. For practical purposes, discharge lines from small Xe, Ne, Ar, Hg, and Kr lamps are good "infinitely" narrow sources. For each spectrometer the operator should prepare a table showing the spectral slit as a function of absolute wavelength and mechanical slit width. Additionally, the rare gas discharge lines are very useful for calibrating the absolute wavenumber or wavelength accuracy of the spectrometer and thus the positions of Raman peaks. When collecting data for publication a calibration line should be run before and after each spectrum. Wavenumber values in air and vacuum differ by several cm^{-1} (see Table 3.1), and care should be taken to ensure that the wavenumber values taken for the laser excitation line and calibration lines are either both in air or both in vacuum.

For each spectral scan it is important to keep a balance between scan speed, spectral slit, and the time constant imposed to smooth out noise. Failure to do this leads to distortion of the spectrum with consequent inaccurate peak positions and poor resolution. A good rule of thumb to set the correct balance is

$$\frac{\text{time constant (sec)} \times \text{scan speed (cm}^{-1}/\text{sec)}}{\text{spectral slit (cm}^{-1})} \leq 0.25 \qquad (3.4)$$

Keeping the foregoing concepts in mind, we can check the resolving power of the instrument. This is the ability to separate closely spaced Raman peaks. The 459-cm^{-1} peak due to the totally symmetric stretch of CCl_4 provides a good test for resolving power. This feature should be partially resolved into components each due to species containing different chlorine isotopes, $CCl_2^{35} Cl_2^{37}$ (455.1 cm^{-1}), $CCl_3^{35} Cl^{37}$ (458.4 cm^{-1}), and CCl_4^{35} (461.5 cm^{-1}). The spectral slit should be set at 1–2 cm^{-1} and the scan rate and time constant set to be in keeping with Eq. (3.4)

The last important spectral parameter dependent on the spectrometer involves Raman peak intensities. Since the Raman effect is intrinsically weak, it is important that the highest possible percentage of the Raman photons passing through the entrance slit reach the detector at the exit slit. There is always a compromise between the light gathering power of a spectrometer and its resolving power. In biochemical studies it is often worthwhile to select a spectrometer which sacrifices resolution for increased light gathering ability. A practical test for the sensitivity of a complete Raman instrument is that, using 100 mW of laser power at 488 nm and 8 cm^{-1} spectral slit, the bending mode of water at 1645 cm^{-1} should be clearly detectable. The 1645-cm^{-1} feature can be seen in the Raman spectrum of water given in Fig. 1.3.

In addition to its overall light gathering efficiency, the relative response at different wavelengths of a Raman instrument, called the *spectral sensitivity*, is needed to measure accurate *excitation profiles*. The latter are often reported in resonance Raman studies and are a measure of the relative intensity of a Raman line as a function of the excitation wavelength. To measure spectral sensitivity, a standard lamp with known spectral output at differing wavelengths replaces the sample. With the collecting optics in place and care taken not to overload the detector, the spectrum of the lamp is then measured using the Raman spectrometer. By correcting the measured spectrum for the known output of the lamp the spectral response of the instrument is obtained. Since the spectrometer response varies with the angle of the E vector of the light entering the instrument, all intensity measurements should be made with the calcite scrambler in place.

II. Sampling Techniques and Sample Problems

A. Sampling Techniques

Regarding the sample, the inherent requirements of a modern Raman experiment are simple. Taking into account the direction of the E vector, a laser beam is focused on, or in, the sample, and the scattered light is analyzed for frequency and intensity. Being a narrow, unidirectional entity, a laser beam can be easily manipulated in a variety of sample-cell configurations. The experimenter can, therefore, exercise considerable ingenuity in the design and use of sample cells. A substantial advantage, stemming from the geometric simplicity of the Raman experiment, is that samples may be examined in any physical state. For liquid samples a cell consisting of a 1-cm path length cuvette, normally used in a fluoresence spectrometer, is adequate, provided, of course, the cell bottom is transparent. The cell should be taped around the meniscus to reduce the amount of scattered light from the interface reaching the spectrometer. Since biological samples can be purified often in only small quantities, it is important to develop a sampling system that is able to utilize microliters of solution. A variety of arrangements based on capillary tubing are available; the one favored in the author's laboratory is shown in Fig. 3.4. The sample is introduced into the tube by a micropipette and "shaken down" with the motion used to set a clinical thermometer. The capillary is sealed by wrapping Parafilm around the open end. Assuming the spectrometer has vertical slits, the capillary is used in the vertical position, which offers the advantage of a long path length for the beam, as well as elimination of scattered light from the air–glass–solution interface since this will be focused off the slit. Moreover, since the laser beam enters from the top, any heat absorbed by the solution rises to the top of the cell and can be dissipated by a stream of cooled air (Fig. 3.4). Capillary cells of the kind shown in Fig. 3.4 can be made by a glassblower from material with an approximately 1.5-mm internal diameter which has been drawn from quartz tubing.

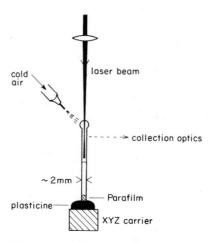

Fig. 3.4. A capillary Raman cell for small quantities of liquids. As little as a few microliters of liquid may be used.

Many candidates for RR spectroscopy are photolabile; i.e., they either photodecompose in the laser beam or, as in the case of the visual pigments, undergo photoinduced transformations that reflect their biological function. To obtain RR spectra from any of these samples, it is necessary to move the sample through the beam at a rate at which the buildup of photo-products becomes insignificant. Two methods, based on a spinning cell and sample flow, are widely used to move the sample through the beam. The former, the rotating Raman cell, was developed by Kiefer and Bernstein (1971) and is depicted in Fig. 3.5. Variants of the original design have been developed, for example, to automatically record depolarization ratios (Kiefer, 1977). About 1 ml of sample is required in the smallest practical rotating cell. At many revolutions per minute the rotating cell acts as a centrifuge, and the deposition of particulate matter (e.g., cells from a bacterial cell suspension) on the outer wall may occur. In addition, for any sample very little mixing occurs at high rotation speeds, and a "burned out" annulus can form in the cell (this is clearly seen if the experiment is monitored on a television screen using "rapid Raman" technology, Section III,C). However, these are minor problems, the rotating cell was an important advance and remains an essential tool. A simple alternative to the rotating cell is to stir the sample solution using a magnetic stirrer. This arrangement is shown schematically in Fig. 3.6; it too has a minimum volume requirement of about 1 ml. The "stirring cell" has proven useful in the author's laboratory for studying kinetic reactions occurring on the time scale of minutes; reactants can be mixed and a Raman scan begun with the minimum of delay.

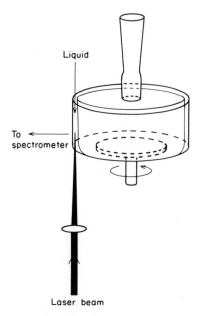

Fig. 3.5. The rotating Raman cell.

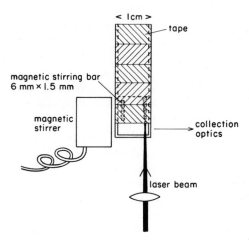

Fig. 3.6. A quartz cuvette used as a "stirred" Raman cell. The cell is taped to prevent light scattered from the meniscus from entering the collection optics.

The second major technique for obviating sample decomposition uses a flow system. In its simplest form the flow technique uses a horizontal quartz capillary tube of 1–2-mm internal diameter with the vertical laser beam focused into the tube. If only a few milliliters of sample are available, a closed loop can be set up with the solution circulated by a peristaltic pump. This has the possible disadvantage of recycling photo-products through the beam. If larger volumes of samples can be used, it is preferable to push the solution once through the capillary cell by means of a syringe pump. Flow speeds of 5–20 cm/sec were found to be sufficient to eliminate photochemical problems in a study of chromophoric enzyme–substrate intermediates excited by 350.7-nm irradiation (MacClement *et al.*, 1981). The syringe pump has the advantage of a smooth action, and between 5 and 20 ml are required to obtain a RR spectrum at the speeds cited. For extremely photolabile samples such as the visual pigments, very high flow speeds, up to 500 cm/sec, are used to quantitatively control photoisomerization (Chapter 5, Section VI). In experiments using the visual pigments, sample volumes of 200 ml were recycled through the beam.

A variety of arrangements are available for solid samples. Single crystals are mounted on a goniometer head for Raman experiments. The individual components of the scattering tensor can be obtained only in single crystal studies and confirmatory information on crystal symmetry may also be forthcoming. Fibers may also be stretched and mounted as oriented samples. Minute amounts of solid sample can be studied using the Raman microprobe (Rosasco, 1980). The microprobe uses a conventional Raman spectrometer but special attention is given to the foreoptics. The microsample is illuminated by a tightly focused laser beam which is theoretically limited in size by diffraction only. The majority of solid samples of a biochemical origin are polycrystalline, and these may be examined by holding them in a capillary tube or by pressing them into pellets. If the sample is not transparent to the laser beam, it will probably need to be rotated in a version of the rotating Raman cell in which the sample is mixed with KBr and pressed into an annulus in a rotatable steel disk (see Kiefer, 1977). Another arrangement which overcomes the problem of sample heating (Koningstein and Gächter, 1973) keeps the sample fixed and "wobbles" the point of illumination by a slightly eccentric motion of a plane glass plate in the path of the laser beam. For any of the aforementioned techniques a few milligrams of sample is usually sufficient.

Raman spectroscopy can also be used effectively to study thin films and surface phenomena generally. The Raman spectrum of a monolayer absorbed at the interface of two solutions can be recorded using the total reflection method of Takenaka and co-workers (Takenaka, 1979).

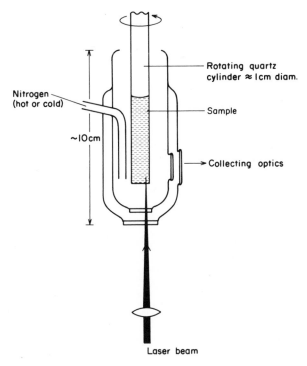

Rotating quartz
cylinder ≈ 1cm diam.

Nitrogen
(hot or cold)

Sample

~10cm

Collecting optics

Laser beam

Fig. 3.7. A simple variable temperature rotating Raman cell with a usable range of at least −60° to +100°C.

For variable temperature experiments, small amounts of photostable sample can be mounted in a thermostatically controlled metal block (Thomas and Barylski, 1970) or in a stream of heated or cooled N_2 gas (Miller and Harney, 1970). The latter technique has a particularly wide temperature range, from −190° to 200°C. Several ingenious variable temperature arrangements encompassing the rotating cell have been suggested. In the author's laboratory a variable temperature rotating cell has been devised which incorporates simplicity, small sample volume, and temperature control over a range suitable for studying biological molecules. The cell and its jacket are shown in Fig. 3.7.

B. Sample Problems

In many cases the key to obtaining good quality Raman or resonance Raman spectra from biochemical systems is sample purity. Sample problems leading to the inability to record spectra can be frustrating and may quickly disillusion a newcomer to the technique. The remedy is simple: "Do not give up after the first attempt!" There are many relatively simple methods to clean up the sample, and with a little experience good spectra soon become the norm.

For liquids or solutions it is first necessary to ensure that they are optically homogeneous. This is best achieved by low-speed centrifugation, for example, in a bench top clinical centrifuge, or passage through a 10–100 micron filter (e.g., Millipore filters). Failure to remove particulate matter from the solution may increase the background attributable to scattered light, and, in stationary cells, cause "hot spots" as particles heat up in the laser beam.

Probably the most often encountered hurdle to acquiring Raman data is background luminescence from fluorescent or phosphorescent processes, or both. Luminescence, being inherently more probable than Raman scattering, often obliterates the weaker Raman signal. Luminescence stems from either of two possible sources: first, from a chromophore which forms a natural part of the system under study, or, second, from a chromophore which is present as an impurity. The former situation may require one of the solutions outlined in Sections III,A, and III,B. However, luminescence from impurities is the more common occurrence, and this often can be eliminated by sample purification. The science of biochemical separation is an expanding and technically innovative field; a whole armory of simple or sophisticated tools can be used to improve sample purity. Techniques such as dialysis and basic column chromatography can be mastered in a few hours while others need special biochemical help. Thus, close cooperation between biochemist and spectroscopist may be critical at this stage. Before extensive sample purification is undertaken, it is worthwhile to attempt to reduce the luminescence by changing the excitation wavelength. For RR spectroscopy a change to an excitation wavelength further into the blue is recommended. The rationale behind this is that the position of the maximum wavelength of luminescence emission is independent of the excitation wavelength whereas the absolute positions of Raman peaks change with excitation wavelength (Chapter 2, Section V). Therefore, by shifting the Raman excitation wavelength further into the blue, the Raman peaks may appear between the exciting line and the onset of fluorescence emission. This method may suffice even in cases where the luminescence is intrinsic to the system being studied. For instance, in the author's laboratory RR spectra could not be obtained from the complex $4NH_2$–$3NO_2$–cinnamoyl–chymotrypsin, using its absorption band near 430 nm, with any excitation line to the red of 460 nm. However, using 441.6-nm excitation, the Raman to luminescence photon ratio was sufficient to give resonance

Raman spectra of acceptable quality. The same approach has also been used to obtain spectra of some flavin compounds which are discussed in Chapter 5, Section VII. For the case of nonresonant spectra "going to the red," e.g., using the 647.1-nm Kr^+ line, often eliminates luminescence problems. A luminescent background can sometimes be reduced by adding a quenching agent to the sample solution. Potassium iodide, at concentrations of 0.2–1.0 M in KI, has been used. Care must be taken that the KI does not chemically modify the sample. Burning off fluorescence by exposing the sample to intense laser irradiation (with power in the watt range) for several hours also has its proponents. However, some biochemists question the integrity of their samples after such treatment.

III. Sophisticated Techniques

A. Coherent Anti-Stokes Raman Spectroscopy

Coherent anti-Stokes Raman spectroscopy (CARS) is of interest for biochemical studies because it offers a general means of obtaining a RR spectrum of a highly fluorescent chromophore. The CARS technique involves the use of two tunable dye lasers set at frequencies ω_1 and ω_2 respectively (Fig. 3.8). If these two light beams cross in the sample at the phase matching angle θ, coherent anti-Stokes emission at $\omega_3 = 2\omega_1 - \omega_2$ is generated through the third-order linear polarization. The laserlike beam caused by ω_3 is greatly enhanced when the frequency interval $\omega_1 - \omega_2 = \Delta$ is equal to a Raman-active molecular vibrational frequency. Thus vibrational Raman spectra are obtained by fixing the frequency of ω_1 and varying the frequency of ω_2. Fluorescence rejection occurs in a CARS experiment because the signal beam ω_3 is spatially and temporally removed from the fluorescence signal.

A major problem in the application of CARS to molecules in aqueous solution is interference from the solvent, which can contribute a background emission resulting from the third-order susceptibility of the solvent. To a certain extent this problem may be overcome when the CARS experiment is conducted under resonance conditions. As the ω_1 beam frequency approaches an electronic transition of the solute, the solute's CARS signal is resonance enhanced whereas the background emission remains unchanged. However, even under resonance conditions, interference from the background can create poorly defined lineshapes (Dutta *et al.*, 1977), and care must be taken to produce reliable results. An early application of resonance CARS to a biochemical system involved the fluorescent isoalloxazine chromophore found in the coenzyme flavin

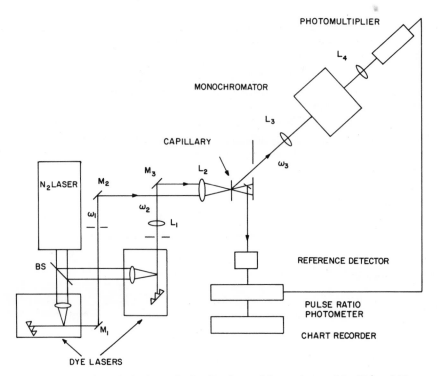

Fig. 3.8. CARS optical schematic. L_1, L_2, L_3, and L_4 are lenses, M_1, M_2, and M_3 are mirrors, and BS is a beam splitter. (Reproduced from Carreira *et al.*, 1977, by permission.)

adenine dinucleotide (FAD) (Dutta *et al.*, 1977). Resonance CARS spectra of isoalloxazine in FAD free in solution and of isoalloxazine riboflavin binding protein are shown in Fig. 3.9. Flavins are discussed in detail in Chapter 5, Section VII; here we simply note that there are important spectral differences, for example, in the $1250-1300$-cm^{-1} region, between the spectrum of the isoalloxazine chromophore in H_2O and in D_2O for both the FAD coenzyme and the riboflavin binding protein.

B. Fluorescence Discrimination by Time Resolution and by Wavelength Modulation

In addition to CARS there have been two further technical innovations capable of discriminating between Raman and fluorescence photons. The first depends on the difference in time scales between the fluorescence process (typically nanoseconds) and the Raman or RR processes (typically picoseconds). The basic concept is to use very short laser pulses and

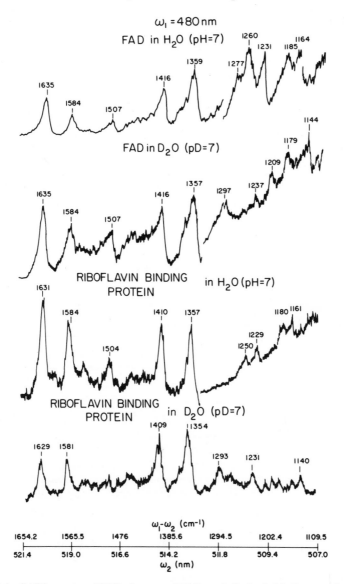

$\omega_1 = 480$ nm

FAD in H_2O (pH=7)

FAD in D_2O (pD=7)

RIBOFLAVIN BINDING PROTEIN in H_2O (pH=7)

RIBOFLAVIN BINDING PROTEIN in D_2O (pD=7)

$\omega_1 - \omega_2$ (cm^{-1})

| 1654.2 | 1565.5 | 1476 | 1385.6 | 1294.5 | 1202.4 | 1109.5 |

| 521.4 | 519.0 | 516.6 | 514.2 | 511.8 | 509.4 | 507.0 |

ω_2 (nm)

Fig. 3.9. CARS spectra of FAD, deuterated FAD, and riboflavin binding protein in H_2O and D_2O; $\omega_1 = 480$ nm and ω_2 scan speed = 0.6 nm/min. (Reproduced from Dutta *et al.*, 1978, by permission.)

to synchronously gate the detector so that the detector is "on" while the Raman photons from a given pulse arrive, but is "off" before the majority of fluorescence photons arrive. The detector is then gated on to meet the Raman photons arriving from the next pulse and so forth, (Van Duyne *et al.*, 1974; Harris *et al.*, 1976; Yaney, 1976).

The second method was developed by Funfschilling and Williams (1976) and utilizes a periodic change (or modulation) in wavelength of the excitation irradiation. The principle underlying the method is that the narrow Raman lines will move, in terms of wavelength, in step with the excitation line whereas the broad fluorescence band will remain essentially constant. By using a phase sensitive lock in amplifier it is then possible to pick out the moving Raman lines against the constant background.

C. Time-Resolved Raman Spectroscopy and the Raman Microscope

Time-resolved or "rapid" Raman spectroscopy refers to the recording of complete Raman spectra in a short time span and spectra have been obtained within nanoseconds, or even picoseconds. In biochemistry time-resolved Raman spectroscopy has obvious applications for characterizing transient species, photobiological processes, and excited electronic states. The experimental outline of a rapid Raman instrument is given in Section I,C and Figure 3.3; the normal sampling arrangement is used in conjunction with a nonscanning grating spectrograph and a multichannel detector. An additional point is that pulsed lasers, with their short high-power bursts of photons, find special utility in the rapid Raman technique, and in experiments demonstrating the potential of this method complete spectra of acetone have been obtained from a single 25-picosecond pulse (Bridoux and Delhaye, 1976).

The principal pioneers in the development of rapid Raman spectroscopy were Bridoux, Delhaye, and co-workers at Lille, France. They have summarized their work in a recent review (Bridoux and Delhaye, 1976), and a schematic of a multichannel spectrometer developed by the French group is shown in Fig. 3.10. Bridoux and Delhaye have outlined the advantages of multichannel detection as follows.

1. Simultaneous analysis of multiple data points; intensifier tubes and television camera tubes allow the analysis of up to 10^5 channels of information.

2. High sensitivity; the sensitivity is similar to that of a good photomultiplier.

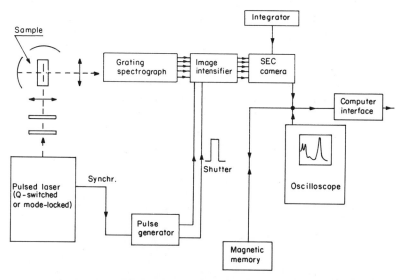

Fig. 3.10. Schematic diagram of a multichannel spectrometer used for time-resolved Raman spectroscopy. (Reproduced from Bridoux and Delhaye, 1976, by permission.)

3. Photometric reproducibility.

4. Storage capacity; this is of the same order of magnitude as that of sensitive photographic material.

5. Short response time; this permits the storage of all the spectral elements in a "single exposure" of duration as short as a few picoseconds.

The short response time allows the detector to be switched on, by electronic gating, for just the duration of a laser pulse. This effectively eliminates the dark current due to thermionic emission of the photocathode since the dark current is negligible during the span of the few nanoseconds "on" time. The gating mechanism is shown between the laser and image intensifier in Fig. 3.10.

Time-resolved Raman spectroscopy is increasingly used in biochemical studies of the visual pigments (Chapter 5, Section VI) and heme proteins (Chapter 5, Section II).

In another application, two groups (Dallinger *et al.*, 1979; Jensen *et al.*, 1980) have reported, the RR spectrum of the lowest triplet excited state of β-carotene. This excited state, which has a half-life of approximately 2 μsec, was generated by pulse radiolysis and the RR spectrum obtained by multichannel detection. The RR spectrum of the ground state of all-trans β-carotene is shown in Fig. 5.20. Table 3.2 compares the RR spectra of the ground and excited states, and the marked differences in

Table 3.2

RR Spectra of β-Carotene

Ground state[a]	Triplet state[a,b]
961 (vw)	965 (w)
1003 (w)	1009 (m)
1157 (vs)	1125 (vs)
1193 (w)	1188
1215 (vw)	1236 (s)
1521 (vs)	1496 (vs)

[a] vw, very weak; w, weak; m, medium; s, strong; vs, very strong.
[b] Taken from Jensen *et al.* (1980), features at 1014, 1126, and 1495 cm^{-1} were reported by Dallinger *et al.* (1979).

the two spectra show that significant electron reorganization and possibly conformational change occur in the triplet state.

The temperature-jump technique, in which a system is rapidly subjected to a change in temperature and then observed to relax back to its equilibrium state, is a powerful means of kinetic analysis. The usual means of monitoring the relaxation process is by absorption or fluorescence spectroscopy. Peticolas and co-workers (see Sturm *et al.*, 1978) have demonstrated that multichannel detection makes temperature-jump Raman spectroscopy feasible. They were able to obtain Raman spectra of a 2% solution of double helical polyadenylic acid and polyuridylic acid (see Chapter 7, Section II) 2 and 100 msec after the temperature jump.

The *Raman microscope* uses a multichannel detector to provide an image of a microscopic object according to the Raman active frequencies of the components making up the object. There are two main classes of multichannel detectors. The first is a diode array which is a linear array of miniscule diodes each of which functions is a miniature photomultiplier tube. The second class is an image intensifier tube which is similar to a low-light-level television camera. Diode arrays can collect information in only one dimension whereas an image intensifier permits collection over the photoactive surface of the intensifier, i.e., in two dimensions. Hence an image intensifier affords an entire two-dimensional picture and gives rise to the potential for performing Raman microscopy. This is another field in which Bridoux, Delhaye, and their co-workers have been very active. In the Raman microscope the laser beam is focused into a sample, and the focal point is transferred by a microscope and a spectrograph to

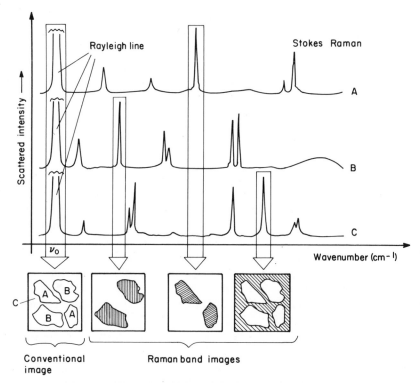

Conventional
image

Raman band images

Fig. 3.11. Illustration of the principle of selective imaging by use of specific spectral "bands" of the Raman-scattered radiation. A and B are inclusions within the matrix C; each material possesses a characteristic Raman spectrum shown as A, B, and C, respectively. (Reproduced from Rosasco (1980) by permission, original due to Dhamelincourt (1979).)

the active surface of an image intensifier. By adjusting the spectrograph it is possible to select different positions of the sample according to their Raman peaks. This principle is shown in Fig. 3.11 in which the laser beam is focused onto the area labeled "conventional image." In a standard Raman experiment the Raman spectrum would contain contributions from substances A and B and the matrix C in which A and B are embedded. However, by tuning the spectrograph within a Raman microscope assembly the image of A or B or C will appear on the television monitor attached to the image intensifier as the Raman active wavenumbers for A or B or C, respectively, are selected.

The Raman microscope has, at present, been used to examine biological samples having little or no fluorescent material within them. The advantage of being able to combine chemical identification and location within a physiological structure is immediately obvious. An interesting example concerns the application to some 3.8×10^9-year-old fossils. Cell-like inclusions, of approximately 30-μm diameter, occurring in cherty layers of quartzite were examined by a Raman microscope (Pflug and Jaeschke-Boyer, 1979) and gave rise to Raman bands characteristic of organic materials. This *in situ* finding was taken as strong evidence that the inclusions contained molecules of biological origin and that these inclusions are indeed microfossils.

D. Raman Optical Activity

Optically active molecules give rise to very small differences in the intensities of Raman scattering from right and left circularly polarized light (Barron, 1978). The correct interpretation of these differences, although not trivial, should provide new insights into conformation and configuration about chiral centers. The major biological example, at present, of the observation of Raman circular intensity differential (CID) is from cytochrome c (Barron, 1975). Attempts at observing Raman CID from unperturbed hemes have, thus far, been unsuccessful. However, Barron (1975) was able to induce CID from ferrocytochrome c by using a magnetic field of approximately 0.7 T. The CIDs are small ($\pm 2 \times 10^{-3}$), but were clearly confirmed by reversing the field polarity or by removing the field. The CIDs are observed in the 1100–1600-cm^{-1} region where the most intense porphyrin modes are observed. Quite different CIDs are observed for 514.5-nm and 501.7-nm excitation (Barron, 1977).

Suggested Review Articles for Further Reading

Barron, L. D. (1978). Raman optical activity. *Adv. Infrared Raman Spectrosc.* **4**, 271.

Bridoux, M., and Delhaye, M. (1976). Time-resolved and space-resolved Raman spectroscopy. *Adv. Infrared Raman Spectrosc.* **2**, 140.

Kiefer, W. (1977). Recent techniques in Raman spectroscopy. *Adv. Infrared Raman Spectrosc.* **3**, 1.

Nibler, J. W., Shaub, W. M., McDonald, J. R., and Harvey, A. B. (1977). Coherent anti-Stokes Raman spectroscopy. *Vib. Spectra Struct.* **6**, 173.

Rosasco, G. J. (1980). Raman microprobe spectroscopy. *Adv. Infrared Raman Spectrosc.* **7**, 223.

CHAPTER 4

Protein Conformation from Raman and Resonance Raman Spectra

I. Protein Structure

Polypeptides and proteins are formed from amino acid building blocks. The general formula for an amino acid is $RCH(NH_3^+)CO_2^-$, where the side chain R can be one of 20 commonly found variants. The formulas of the R groups and the pKs of the amino acid side chains are given in Table 4.1. The state of ionization of amino acid side chains within a protein often has important consequences for the structure and function of the entire molecule. For an introduction to the vibrational spectroscopy of amino acids themselves, consult Parker's (1971) monograph since this topic is not discussed in this volume.

Formally, two amino acids may be linked together by elimination of a water molecule to form a dipeptide (Fig. 4.1). Polypeptides containing two or more amino acids are found in nature as degradation products of proteins and, in addition, as functional units possessing important biological activity. An outstanding example of the latter are the recently discovered enkephalins; these are pentapeptides found in the brain which possess opiate properties. Synthetic homogeneous polypeptides containing a single repeating amino acid unit have been very useful in establishing the structure–spectra correlations needed to interpret the Raman spectra of the more complex heterogeneous assemblies of peptide units in proteins. The main constituent of proteins is an unbranched polypeptide chain, or

Table 4.1

The Common Amino Acids

Amino acid	Side chain, R $(RCH(NH_3^+)CO_2^-)$	pK_a of side chain
Glycine	H—	—
Alanine	CH_3—	—
Valine	$\begin{array}{l}H_3C\\ \diagdown\\ CH—\\ \diagup\\ H_3C\end{array}$	—
Leucine	$\begin{array}{l}H_3C\\ \diagdown\\ CHCH_2—\\ \diagup\\ H_3C\end{array}$	—
Isoleucine	$\begin{array}{l}CH_3H_2C\\ \diagdown\\ CH—\\ \diagup\\ H_3C\end{array}$	—
Phenylalanine	⬡—CH_2—	—
Tyrosine	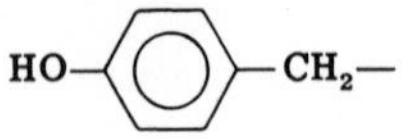 HO—⬡—CH_2—	10.1
Tryptophan	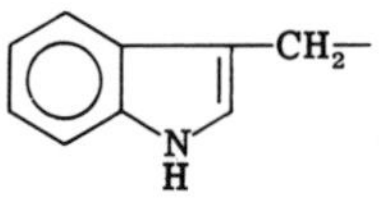 indole—CH_2—	—
Serine	$HOCH_2$—	—
Threonine	$\begin{array}{l}HO\\ \diagdown\\ CH\\ \diagup\\ H_3C\end{array}$	—
Cysteine	$HSCH_2$—	8.4
Methionine	$CH_3SCH_2CH_2$—	—
Asparagine	$H_2NC(=O)CH_2$—	—
Glutamine	$H_2NC(=O)CH_2CH_2$—	—
Aspartic acid	$^-O_2CCH_2$—	3.9
Glutamic acid	$^-O_2CCH_2CH_2$—	4.1
Lysine	$H_3N^+(CH_2)_4$—	10.8
Arginine	$\begin{array}{l}H_2N^+\\ \diagdown\\ C—NH(CH_2)_3—\\ \diagup\\ H_2N\end{array}$	12.5
Histidine	$\begin{array}{l}HC{=\!=}C—CH_2—\\ ||\\ NNH\\ \diagdown\diagup\\ C\\ |\\ H\end{array}$	6.0
Proline	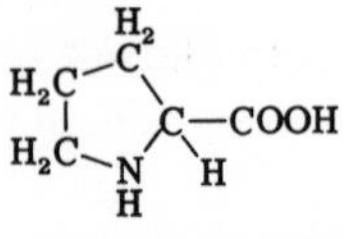 imidazolidine ring with —COOH	—

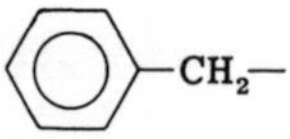

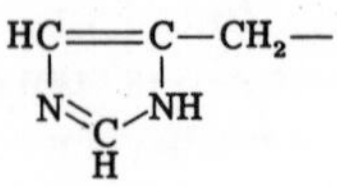

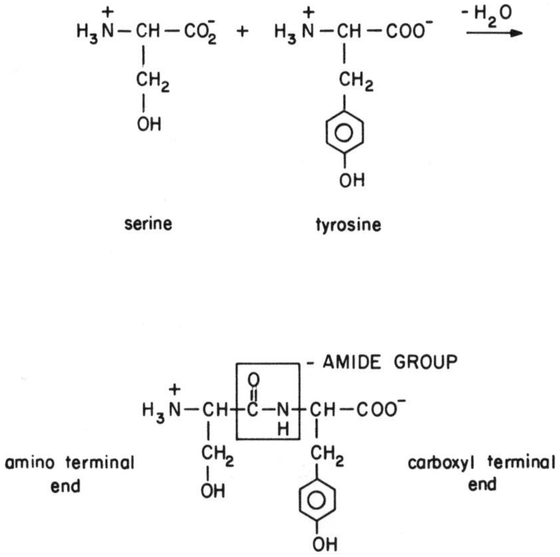

$$H_3\overset{+}{N}-CH-CO_2^- \;+\; H_3\overset{+}{N}-CH-COO^- \;\xrightarrow{-H_2O}$$

serine   tyrosine

AMIDE GROUP

$$H_3\overset{+}{N}-CH-C-N-CH-COO^-$$

amino terminal end   carboxyl terminal end

Fig. 4.1. Linkage of amino acids to form a dipeptide.

chains, consisting of L-α-amino acids linked by amide bonds between the α-carboxyl of one residue and the α-amino group of the next. The *primary* structure of a protein is defined by the amino acid sequence.

The *secondary* structure of a protein is the hydrogen-bonded helical or zigzag arrangement of polypeptide chains along the long axis in fibrous proteins. The latter are insoluble and serve as structural elements in materials such as silk and hair fibers. The parallel and relatively extended peptide chains making up the secondary structure give rise to the rod or sheetlike properties of fibrous proteins. A common helical arrangement found in proteins is the right-handed α-helix shown schematically in Fig. 4.2. Within the helix, each amide group makes a hydrogen bond with the third amide group away from it in either direction. The C=O groups are parallel to the axis of the helix and are almost in a straight line with the NH groups to which they are hydrogen bonded. The R side chains of the amino acids point away from the axis. There are approximately 3.6 amino acids in each turn of the helix.

An extended polypeptide chain can form hydrogen bonds to a second extended polypeptide chain running parallel or antiparallel to the first. The second chain can, in turn, make additional complementary hydrogen bonds to a third chain to form a sheet. When all the chains run in the same direction, a parallel β-pleated sheet is formed; when the chains alternate directions, the antiparallel β-pleated sheet results. Figure 4.3 shows two chains binding to form part of an antiparallel β-sheet.

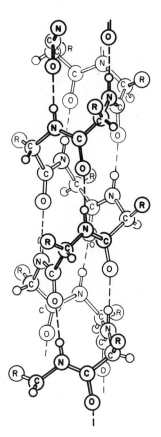

Fig. 4.2. A right-handed α-helix. (Adapted from Linus Pauling: *The Nature of the Chemical Bond, Third Edition.* Copyright © 1960 by Cornell University. Used by permission of the publisher, Cornell University Press.

Tertiary structure refers to the way in which the polypeptide chains are folded within globular proteins. The latter have tightly folded polypeptide chains and are roughly spherical or egg shaped. The globular proteins have a dynamic role and are the "workers" in biochemistry; for example, they perform catalytic and transport functions. Globular proteins also contain domains possessing helical or β-sheet order, but often contain large regions in which the peptide chain appears to be in a random conformation. An additional structural feature of protein conformation is the β-turn involving four consecutive residues where the polypeptide chain folds back on itself by nearly 180°. These turns have also been called β-bends, hairpin loops, reverse turns, and 3_{10} bends. They have been

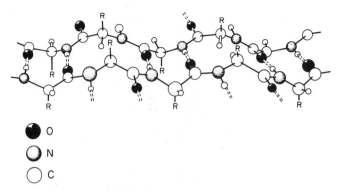

increasingly recognized as important components of protein structure. Globular proteins may contain one or more polypeptide chains and, in addition, these chains may be cross-linked by covalent –S–S– bridges from two cysteine residues joined by their thiols. In this way cross-linking can occur within a single chain or between separate polypeptide chains. Many proteins also have a prosthetic group bound to the polypeptide structure; e.g., heme proteins contain the porphyrin moiety, and metalloproteins have one or more metal ions bound to peptide groups.

The three-dimensional conformation of a protein molecule is determined by its amino acid sequence. Proteins may be denatured, i.e., unfolded, by changing temperature or pH or by, for example, the addition of detergent to a solution of globular proteins. Denaturation causes proteins to lose their biological activity, but in some cases denaturation is fully reversible.

Quarternary structure denotes the manner in which association takes place between individual polypeptide chains. For example, the quaternary structure of the enzyme glyceraldehyde-3-phosphate dehydrogenase consists of a particle composed of four monomers, each consisting of a single polypeptide chain of MW 40,000 and having its own catalytic site. The native enzyme thus contains four catalytic sites, and this offers the potential for the control of activity through the use of one site affecting the activity of the three remaining sites. In general, the monomers can be dissociated and associated by varying conditions such as pH or salt content in solution since, usually, the monomers are held together by noncovalent forces. In fibrous proteins, the term quaternary structure is sometimes used to describe the way in which, for example, helices are packed to make up the fibers.

II. Introduction to Raman and Resonance Raman Studies of Proteins

Proteins and their components have the distinction of being the classical example of the application of Raman spectroscopy to biomolecules. Edsall and co-workers studied in detail the effects of ionization and of deuterium substitution on the Raman spectra of amino acids and related compounds (Edsall, 1936) and also obtained the first Raman spectrum of a protein, lysozyme (Garfinkel and Edsall, 1958). Edsall's work was performed using lamps as excitation sources whereas all present-day studies utilize lasers. To a reasonable approximation recent Raman and RR studies of proteins can be separated along structure and function lines, respectively. The normal Raman spectrum of a protein probes structural details such as average peptide backbone conformation. Although some amino-acid side chains, particularly those of tyrosine and tryptophan, can also be probed, it is often not possible to focus on a single molecular group in the protein. In contrast, most RR studies focus on a single chromophore, and since this is often associated with the biochemical role of the protein, the RR spectrum provides a probe of function.

In this chapter the normal Raman spectra of proteins, which are obtained usually by excitation in the 450–650-nm region, will be discussed first. Then brief mention will be made of the potential of UV (below 300 nm) excited RR spectra of proteins where aromatic amino acid side chains and the peptide linkages themselves are the chromophores of interest. In Chapter 5 the extensive amount of RR work on naturally occurring protein-associated chromophores, absorbing in the 350–650-nm range, will be outlined and this will be followed in Chapter 6 by a description of the RR labeling technique. In the latter, extrinsic chromophores, chosen to mimic natural biological components, are placed at protein sites of interest as RR probes or reporter groups.

III. Raman Studies of Proteins

A. *Characteristic Frequencies of the Different Backbone Conformations in Synthetic Polypeptides*

The observed Raman spectrum of a protein recorded under nonresonance conditions consists of contributions from various amino acid side-chain modes together with modes originating from the peptide backbone. Among the latter, the amide I and amide III modes are widely used to

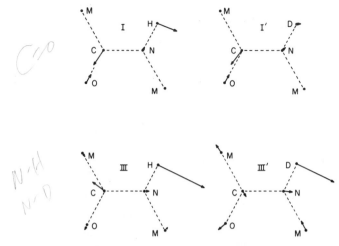

Fig. 4.4. Calculated normal modes of the amide I and III vibrations of *N*-methylacetamide and of the amide I' and III' vibrations of *N*-deuterated *N*-methylacetamide. (Reproduced from Miyazawa *et al.*, 1958, by permission.)

characterize the secondary structure of the peptide backbone. The normal modes of the amide I and III vibrations calculated for *N*-methylacetamide are shown in Fig. 4.4 along with the amide I' and III' modes in which the N–H moiety is replaced by N–D. The amide I and I' modes have a high degree of C=O stretching character while the amide III and III' modes have a large contribution from N–H (or N–D) in-plane bending. The ranges within which the amide I and amide III Raman bands occur for α-helical, β-sheet, and unordered structures are given in Table 4.2. Addi-

Table 4.2

Approximate Positions (cm⁻¹) of the Most Intense Amide I and III Bands in Raman Spectra and Amide I Band in Infrared Spectra for Various Polypeptide Conformations

	Amide I		Amide III
	Raman	Infrared	Raman
α-Helix	1645–1660	1650	1265–1300^a
β-Sheet	1665–1680	1632	1230–1240
Unordered	1660–1670	1658	1240–1260

^{a} Weak and may be confused with side-chain modes.

tionally, Table 4.2 compares the position of the infrared active amide I modes for these structures.

The amide modes of highly ordered polypeptides and proteins may be split as the result of inter and intramolecular interactions. The basic approach to account for the observed splittings in the amide modes is based on perturbation theory and was developed by Miyazawa (1960). Miyazawa's analysis assumes that the observed frequencies can be equated to an unperturbed frequency plus the contribution of inter and intramolecular interaction terms. Krimm and co-workers (see Bandekar and Krimm, 1979, and references therein) have subsequently presented evidence that the physical origin of the perturbation lies in the coupling of the amide groups' transition dipoles.

The use of the amide I and III regions in the Raman spectrum to characterize the secondary structure of a protein depends on the determination of characteristic frequencies for helical, β-sheet, β-turn, and random protein conformations. This is commonly accomplished by using polypeptide models and proteins of known conformation.

Polypeptide models

The different frequencies in the amide I and III regions for α-helical, antiparallel β-pleated sheet, and random-coiled conformations for polypeptides of known structure are summarized in Table 4.3. Intense bands in the 900–1000-cm^{-1} region, attributed to C_α–C–N stretching, have also been found to be conformationally sensitive. From Table 4.3, we see that the amide I frequencies in the Raman spectra for α-helical structures range from 1645 to 1655 cm^{-1} whereas those of antiparallel β-pleated sheets are distinctly higher at 1666–1679 cm^{-1}. Poly-β-benzyl-L-aspartate is exceptional in that in the α-helical form it gives rise to an amide I feature at 1663 cm^{-1}, but this compound's amide I band attributable to the β-pleated sheet conformation is also at the high end of the β-sheet range. In solution, the H$_2$O bending mode near 1645 cm^{-1} often obscures the amide I vibration. However, this problem can be overcome by using D$_2$O as a solvent since the D$_2$O bending mode occurs near 1180 cm^{-1}.

In the amide III region, an antiparallel β-pleated sheet structure gives rise to a band at 1229–1240 cm^{-1} and a weaker feature at 1289–1295 cm^{-1}. Problems arise in the amide III range from overlap of bands from amino acid side chains and from the lack of a well-defined, agreed upon, amide III feature associated with the α-helical structure. The assigned amide III frequencies of α-helices in Table 4.3 show marked irregularity. Nevertheless, it has been suggested that the *absence* of spectral intensity at 1235–1240 cm^{-1} is diagnostic of an α-helix (Lippert *et al.*, 1976). An additional identification band for α-helical structure is a strong skeletal vibration

Table 4.3

Peptide-Backbone Raman-Active Vibrations of Polypeptides of Known Structures[*]

Polypeptide	Amide I[a] (cm^{-1})	Amide III[a] (cm^{-1})	Skeletal[a] (cm^{-1})
α-Helix (left-handed)			
Poly-L-alanine, solid[b,c]	1655	1265, 1275, 1283	909
Poly-γ-benzyl-L-glutamate, solid[b,d]	1650	1294	934
Poly-β-benzyl-L-aspartate, solid[e]	1663	—	890
Poly-L-glutamic acid, solid[f]	1652	1290 (w)	926
Poly-L-leucine, solid[c,d]	1653	1294, 1261	931
Poly-L-lysine			
HCl, solid at 50% humidity[b,g]	1655	1295	—
HCl, film at 92% humidity[h]	1647	1256, 1218	945
HPO$_4$, single crystal[h]	1645	1246, 1210	—
Aqueous solution, pH 11.8, 4°C[i,j,k]	1645	1311 (s), 1200–1300 (vw)[l]	945
Antiparallel β-pleated sheet			
Polyglycine I, solid[p]	1674	1295, 1234	—
Poly-L-valine, solid[b,c,d,m]	1666	1229	—
Poly-L-serine, solid[d]	1668	1235	—
Poly-L-alanine, mechanically deformed solid[c]	1669	1243, 1231[n]	—
Poly-L-lysine, aqueous solution, pH 12, 52°C[i,j]	1670	1240	1002 (w)
Poly-β-benzyl-L-aspartate, heat-treated solid[e]	1679	1237	—
Poly-L-glutamic acid, β$_1$ form[f]	1672	1236	—
Poly-L-glutamic acid, β$_2$ form[f]	1642	1230, 1280	—
Random coil			
Poly-L-lysine, aqueous solution, pH 4[b,i,k]	1665 (br)	1243–1248	958
Sodium poly-L-glutamate, solid[o]	1649	1247	938
Poly-L-glutamatic acid, aqueous solution, pH 11[o]	1656	1249	949

[*] Adapted from Carey and Salares (1980) with permission.

[a] w is weak; vw, very weak; s, strong; br, broad.

[b] Chen and Lord (1974).   [c] Frushour and Koenig (1974a).

[d] Koenig and Sutton (1971).   [e] Frushour and Koenig (1975a).

[f] Fasman et al. (1978a).   [g] Koenig and Sutton (1970).

[h] Yu and Peticolas (1974).   [i] T. J. Yu et al. (1973).

[j] Painter and Koenig (1976b).   [k] Lippert et al. (1976).

[l] Considerable difficulty in assigning the amide III vibration of poly-L-lysine arises from overlapping side-chain vibrations. Low intensity between 1200 and 1300 cm^{-1} has been assigned to the amide III mode[h,i,j] and the strong 1311 cm^{-1} band to lysine side chain vibration.[h] On the other hand, Chen and Lord,[b] and Lippert et al.[k] assign the 1311 cm^{-1} to predominantly amide III vibration.

[m] Fasman et al. (1978b).

[n] Assigned to disordered poly-L-alanine.

[o] Koenig and Frushour (1972a).   [p] Small et al. (1970).

between 909 and 945 cm^{-1} whose intensity markedly decreases following the transformation from α-helix to β-sheet. Unfortunately, this band does not uniquely identify an α-helix since random-coiled polypeptides also have a band with appreciable intensity near 950 cm^{-1}.

For random-coiled polypeptides in solution the amide I and III frequencies fall near 1660 and 1248 cm^{-1}, respectively, and are broader than those of α-helix and β-structures. It is questionable if the frequencies found for the random-coiled polypeptides can be applied to disordered proteins (Hsu et al., 1976; Van Wart and Scheraga, 1978). Characteristic frequencies cannot be expected for unordered polypeptide chains or unordered proteins which consist of a large number of different conformations. A further point to consider in using polypeptides as models for protein structure is the presence of an additional symmetry in the form of repeating side chains in polypeptides which is absent in proteins (Lord, 1977). However, Painter and Coleman (1978) have suggested that some of these difficulties may be overcome by treating unordered polypeptides as a sequence which is distorted from a polypeptide possessing similar dihedral angles within the peptide backbone. By including the effect of transition dipole coupling into the calculations they were able to account for the general features of the amide I mode in the Raman and IR spectra of unordered polypeptides.

Model peptide compounds displaying β-turns have only been characterized recently. Bandekar and Krimm (1979) have performed a vibrational analysis of β-turns and predict that they give rise to amide I frequencies near 1687 cm^{-1} and 1662 or 1642 cm^{-1}, depending on the type of β-turn. These are all expected to be active in the Raman and infrared but the relative intensities of these features in the Raman spectrum is presently unclear. Bandekar and Krimm also predict that the amide III modes of β-turns will appear in the 1290–1330-cm^{-1} region.

B. Determination of Protein Peptide Chain Conformation from the Amide I and Amide III Regions

To assess the suitability of the frequency–structure correlations developed for the homopolypeptides for determining protein structure, the frequencies found for proteins of known structure are given in Table 4.4. The amide I frequency assigned to the α-helical fraction of crystalline glucagon (1658 cm^{-1}) and tropomyosin (1655 cm^{-1}), two highly helical proteins, is close to the observed frequency of the helical polypeptides (1645–1656 cm^{-1}), but in insulin the frequency (1662 cm^{-1}) is higher. Wide variation is found in the amide I frequency of the random fraction of glucagon and insulin (1685 cm^{-1}), feather keratin (1654 cm^{-1}), and α-casein (1668 cm^{-1}).

Table 4.4

Observed Raman Bands for Proteins of Known Structure*

Protein	Structure	Amide I (bands[a]/cm⁻¹)	Amide III (bands[a]/cm⁻¹)	Reference
Glucagon	X-ray: crystals—75% α, 25% random[b] fresh solution—random[c] gel—β[d]	1658 c; 1685 (sh) random	1266 α; 1235 (sh) random	Yu and Liu (1972)
Insulin	X-ray: crystals—52% α, 6% β, 42% random[e] Infrared dichroism: denatured solid—β[f]	— 1672 β 1662 α; 1685 random 1673 β	1248 random 1232 β 1303, 1284, 1269 α; 1240 random 1252, 1227 β	Yu et al. (1972, 1974)
Lysozyme	X-ray: 29–42% α, 10% β, 48–62% random[g]	1660	1272 α; 1258 random; 1238 β	Lord and Yu (1970), Yu and Jo (1973a)
Tropomyosin	various: highly α-helical[h]	1655	1200–1270 (w)	Frushour and Koenig (1974b)
Carboxypeptidase A	X-ray: 30% α, large fractions of parallel β[i]	—	1270 α; 1247 parallel β and random	Yu and Jo (1973b)
Basic pancreatic trypsin inhibitor	X-ray: 25% α, 45% β, 30% random[j]	1667	1289 α, 1265 random; 1242 β	Brunner and Holz (1975)
Ribonuclease A	X-ray: ~18% α, 35–40% β, intermediate geometry[k]	1669	1265 α; 1239 β	Yu and Jo (1973b), Koenig and Frushour (1972b)
Myosin	various: tail > 95% α, Head ~35% α[l]	1650 α	1304 α (fibrous); 1265 α (globular); 1244 β and random	Carew et al. (1975)
Human carbonic anhydrase B	X-ray: 17% helix, 40% β (8% parallel, 32% antiparallel) 43% disordered[m]	1666 β with broad underlying band for disordered	1276 (w, sh) α; 1240, 1231 β; 1258 3_{10} helix	Craig and Gaber (1977)

(continues)

Table 4.4 (Continued)

Protein	Structure	Amide I (bands[a]/cm^{-1})	Amide III (bands[a]/cm^{-1})	Reference
α-Chymotrypsin	X-ray: low α, large amount of twisted β and random[n]	1654 (sh) α; 1668 β	1258 α; 1243	Yu and Peticolas (1974)
Chymotrypsinogen	X-ray: 10–20% α-helix[o]	1668 β	1241 (br) β	Koenig and Frushour (1972b)
Pepsin	X-ray: small α, mostly β and random[p]	1667	1246	Lippert et al. (1976)
Concanavalin	X-ray: predominantly β, some random[q]	1672 β	1242	Painter and Koenig (1975)
Human immunoglobulin G	X-ray and circular dichroism: high β[r]	1673 β	1239 β	Painter and Koenig (1975), Pézolet et al. (1976)
Feather keratin	X-ray and infrared of native: contains antiparallel β[s] Optical rotary disperson and circular dichroism of solubilized: unordered[t]	1671 β	1235–1239	Hsu et al. (1976)
Bovine serum albumin	Optical rotary dispersion solution—55% α, 45% random[u]	1654 1652 α; ~1665 (br, sh) random	1246 1272 α; 1248 random	Lin and Koenig (1976), Chen and Lord (1976b)
α-Lactalbumin	Optical rotary dispersion and circular dichroism: solution—same as lysozyme[v]	1660	1274 (sh) α; 1260 random; 1238 (sh) β	Yu (1974)
β-Lactoglobulin	Optical rotary dispersion and circular dichroism: solution—10–20% α, ~50% disordered, 30–40% β[w]	1662 in solid	1264 (sh); 1243	Bellocq et al. (1972), Frushour and Koenig (1975b)
Ovalbumin	Optical rotary dispersion: solution—25% α, 25% β[x]	1665 in solid	1235 (w) β; 1247 (s) extended β; 1255 irregular structure	Painter and Koenig (1976c)

Ovomucoid

	Optical rotary dispersion: solution—22% α, remainder irregular structure[v]	Optical rotary dispersion: solution—disordered[z]	X-ray: 50% β-sheet[aa]
α-Casein	1665 irregular 1668 in solid	1248 irregular 1254 disordered	Painter and Koenig (1976c) Frushour and Koenig (1974b)
Bence–Jones proteins	1670–1675	1242–1246	Kitagawa et al. (1979c)

* Adapted from Carey and Salares (1980) with permission.
[a] w is weak; s, strong; sh, shoulder; br, broad.
[b] Sasaki et al. (1975).
[c] Gratzer et al. (1968).
[d] Gratzer et al. (1967).
[e] Blundell et al. (1972).
[f] Ambrose and Elliott (1951).
[g] Phillips (1967).
[h] Caspar et al. (1969).
[i] Ludwig et al. (1967); Lipscomb et al. (1968).
[j] Huber et al. (1970).
[k] Kartha et al. (1967).
[l] Lowey et al. (1969).
[m] Kannan et al. (1975).
[n] Blow (1976).
[o] Kraut (1971).
[p] Andreeva et al. (1970); see also Tang et al. (1978).
[q] Edelman et al. (1972).
[r] Poljak et al. (1974); Litman et al. (1973).
[s] Fraser et al. (1971).
[t] Westover et al. (1962); Tiffany and Krim (1969).
[u] Schechter and Blout (1964).
[v] Kronman (1968); Aune (1968).
[w] Roels et al. (1971); Townend et al. (1967).
[x] Chi (1970).
[y] Tomimatsu and Gaffield (1965).
[z] Fasman et al. (1970).
[aa] Epp et al. (1974).

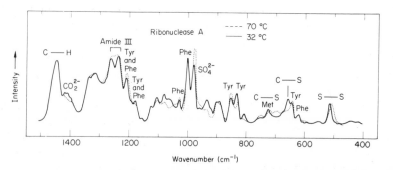

Fig. 4.5. Raman spectra of native and denatured ribonuclease A, at 32° and 70°C, after correction for the water background and being normalized to the intensity of the methylene deformation mode at 1447 cm^{-1}. Protein concentrations of about 10% were used, with typical spectral conditions of 488-nm excitation, 200-mW power, and 7-cm^{-1} spectral slit. Reprinted with permission from Chen and Lord, *Biochemistry* **15**, 1889 (1976). Copyright 1976 American Chemical Society.

The amide I frequencies assigned to the β fractions of the proteins are similar to those found for the polypeptides; however, bands attributable to α or disordered fractions are usually not resolved. The amide III region appears to be more sensitive to heterogeneous secondary structure. Although the amide I regions of lysozyme, basic pancreatic trypsin inhibitor, myosin (Table 4.4), ribonuclease A (Fig. 4.5), and several histone proteins (Guillot *et al.*, 1977) show a single band, the amide III region is split into two or more components. In ribonuclease A the band at 1239 cm^{-1} (Fig. 4.5) is assigned by common consent to the β fraction, but as in the case of polypeptides, the assignment of an amide III contribution from the α-helical fraction is not agreed on. There is an appreciable variation in the amide III frequencies assigned to disordered protein (1235–1270 cm^{-1}), reinforcing the suggestion that there are no characteristic frequencies for unordered regions of proteins. It is also difficult to separate the amide III frequencies of β-sheet structures from those of disordered fractions because of the overlap of bands.

It is apparent that the assessment of protein secondary structure from the amide I and III regions of the Raman spectrum requires a careful analysis taking into account the limitations outlined previously. Considerable progress has been made, however, and Raman spectroscopy retains the unparalleled advantage of being able to monitor protein conformation in the solid, crystalline, and solution phases. Three methods have been put forward in the literature to quantitate the various amounts of secondary structure from the Raman spectrum. In one of these, Pézolet and co-workers (1976) have proposed a means of estimating the amount of

β-sheet structure by using the relative intensity of the peaks at 1240 and 1450 cm^{-1}, which are attributable to the amide III and CH$_2$ bending modes (from the many CH$_2$ groups in the protein), respectively. The method is only applicable when there is a well-defined amide III band at 1240 ± 3 cm^{-1}. The 1450-cm^{-1} feature is used, in Pézolet's approach, as an internal intensity standard, and this idea is common to the second method, developed by Lippert *et al.* (1976), where poly-L-lysine is used to set up standard spectral intensities for α-helical, β-pleated sheet, and random conformations at 1240 cm^{-1} in H$_2$O. Poly-L-lysine, along with the proteins lysozyme and ribonuclease A is also used to "calibrate" intensities for these three conformations at 1632 and 1660 cm^{-1} in the amide I′ region in D$_2$O. The relationship between conformational content and the relative spectral intensities in the Raman spectrum is then given by four simultaneous equations

$$C^{\text{prot}} I^{\text{prot}}_{1240} = f_\alpha I^\alpha_{1240} + f_\beta I^\beta_{1240} + f_{\text{R}} I^{\text{R}}_{1240}$$

$$C^{\text{prot}} I^{\text{prot}}_{1632} = f_\alpha I^\alpha_{1632} + f_\beta I^\beta_{1632} + f_{\text{R}} I^{\text{R}}_{1632}$$

$$C^{\text{prot}} I^{\text{prot}}_{1660} = f_\alpha I^\alpha_{1660} + f_\beta I^\beta_{1660} + f_{\text{R}} I^{\text{R}}_{1660}$$

$$1.0 = f_\alpha + f_\beta + f_{\text{R}}$$

In these equations I^{prot}_{1240} is the height of the protein Raman spectrum at 1240 cm^{-1} in water relative to the spectral height of the methylene bending mode at 1448 cm^{-1}; I^α_{1240} is the "standard" value for 100% α-helical protein; f_α, f_β, f_{R} are the fractions of residues in the α-helical, β-sheet, and random-coil conformations in the protein, respectively; and C^{prot} is a scaling constant which represents the relative methylene band intensity. Using this model Lippert and co-workers were able to derive the percentage of α-helical, β-pleated sheet, and random-coil conformations in several proteins and found good agreement with the results derived by using other techniques.

The third method, proposed by Williams *et al.* (1980a,b), for delineating the secondary structure of proteins is a general extension of the foregoing methods. These workers derive a "standard" intensity for each of several secondary structures at a given frequency and then fit the amide I or amide III′ region of the Raman spectrum of a protein whose structure is sought by a linear combination of the "standards." At a fixed frequency, the observed Raman scattering intensity for a single protein may be expressed

$$f_{\text{ho}} I_{\text{ho}} + f_{\text{hd}} I_{\text{hd}} + f_{\beta\text{a}} I_{\beta\text{a}} + f_{\beta\text{p}} I_{\beta\text{p}} + f_{\text{t}} I_{\text{t}} + f_{\text{u}} I_{\text{u}} = I_{\text{e}}$$

where I_{e} is the normalized experimental Raman intensity, each f is the fraction of a type of secondary structure and I is the intensity attributable to a polypeptide with 100% of the indicated structure type. In the nomenclature of Williams *et al.*, ho stands for ordered helix; hd, disordered helix; βa, antiparallel β-sheet; βp, parallel β-sheet; t, turn; and u, undefined. Normalization is achieved by dividing the observed intensity at each wavenumber by the sum over all wavenumbers of the observed intensities. For the experiments on the amide I profile, six reference intensities I, representing the spectra of polypeptides with a single structure type, were computed from 11 equations, each representing a different protein of known conformation. This set of solutions provided the reference spectra. Linear combinations of the reference spectra were then fitted to the amide I spectrum of the protein for which an estimate of the structure was desired. Good agreement between structures determined by this method and those determined by X-ray analysis was observed. This treatment of Williams *et al.* realizes the power of computer-assisted data processing and spectral fitting and, as an additional advantage, includes structures such as turns which had heretofore been omitted from the analysis. As an interesting aside Williams *et al.* (1980b) noted that the amide III' feature for the α-helical structure in D_2O is intense compared to the analogous feature in the amide III region in H_2O.

Prior to 1979 the characteristic frequencies for the β-turn conformation of protein had not been identified. In the previous section it was noted that Krimm's group, on the basis of calculations, predicted amide I frequencies of 1662, 1642, or 1687 cm^{-1} depending on the type of β-turn, and amide III modes in the 1290–1330 cm^{-1} region (Bandekar and Krimm, 1979). One problem in characterizing β-turns is the lack of well-defined model compounds. However, Bandekar and Krimm (1980) used the approach of calculating the normal modes of the four β-turns in insulin for which precise geometric information, based on X-ray crystallographic studies, is known. Frequencies were predicted in the amide I region near 1652 and 1680 cm^{-1} and in the amide III region above 1300 cm^{-1}. The 1652-cm^{-1} feature overlaps the α-helix band seen in insulin's Raman spectrum at 1658 cm^{-1}, and the 1680-cm^{-1} and amide III bands provide explanations for the features observed at 1681 and 1303 cm^{-1}. Tu and coworkers from their work on various hormones and toxins favor amide I and III modes near 1666 and 1269 cm^{-1}, respectively, for the β-turns in these structures (Bailey *et al.*, 1979).

C. Amide Groups in Nonprotein Structures

Amide modes, especially the amide I vibration, provide monitors for the behavior of amide groups that occur in several important nonprotein environments. A case in point concerns the coenzymes nicotinamide

adenine dinucleotide (NAD⁺) and β-dihydronicotinamide adenine dinucleotide (NADH) which contain the structures

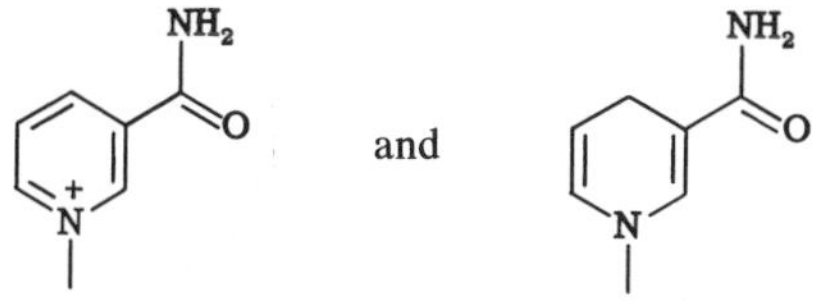

and

respectively. Patrick *et al.* (1974) and Forrest (1976) have shown how the
amide I vibration can provide insight into the differences in the charge
distribution within the amide group in the above structures and have
indicated how the charge distribution may affect the binding to the enzyme. Another study of peptide-like linkages concerns the antibiotic valinomycin (Asher *et al.*, 1977b; Rothschild *et al.*, 1977) which selectively
complexes with alkali cations and facilitates ion transport across membranes. Valinomycin is a 12-membered macrocyclic depsipeptide which
contains alternate peptide and ester linkages. Asher *et al.* (1977b) used the
amide and ester $C{=}O$ peaks to study the conformation of tne uncom-
plexed antibiotic crystallized from several solvents and, relying mainly on
the same spectral region, went on to probe valinomycin conformation in
various polar and nonpolar solvents (Rothschild *et al.*, 1977). The same
group of workers has also used Raman spectroscopy to follow the consequences of cation antibiotic complexation. One detailed study (Asher
et al., 1977a) concerns the nactins, a family of macrocyclic antibiotics,
which, again, are used to study ion transport. The nactins do not contain
amide groups, but they do have four ester linkages which help compose
the ring of the macrocycle. Characteristic Raman spectral changes were
observed for the nactins on cation complexation and for the K⁺, Rb⁺, Cs⁺,
NH_3OH^+, and $C(NH_2)_3^+$ complexes, the ester $C{=}O$ stretch was found to
be linearly proportional to the calculated cation–carbonyl electrostatic
interaction energy.

D. Low-Frequency Vibrations of Proteins

The dynamics of a protein molecule, such as the vibrations affecting
the relative positions of subunits, are thought to play an important role in
biological function. It has been proposed that the low-frequency modes
observed in the Raman and IR spectra of proteins are related to such
overall dynamics. Brown *et al.* (1972a) made the original observation of a
low-frequency mode in the 25–30-cm⁻¹ region of the Raman spectra for
chymotrypsin and pepsin. Genzel *et al.* (1976) also detected a band in this
region in the spectrum of solid lysozyme, but reported that the band could
not be observed in an aqueous solution of the protein. A feature near 25

cm^{-1} in the far-IR spectrum of crystalline lysozyme has also been reported recently (Ataka and Tanaka, 1979). Painter and Mosher (1979) have noted that the low-frequency modes of an antibody molecule are similar to those of lysozyme in that they are considerably diminished in intensity after hydration.

The origin of these low-frequency modes has been separately attributed to either interprotein vibrations or to modes within a single protein backbone. Peticolas (1979) gives a comprehensive account of this interesting and controversial topic.

E. Side-Chain Vibrations

1. The disulfide and –SH groups

A relatively intense Raman line near 500–550 cm^{-1} is observed from ν_{S-S} of a disulfide linkage CCS–SCC; this feature can be seen in the spectrum of ribonuclease shown in Fig. 4.5. Two correlations have been proposed for relating the S–S stretching frequency to conformation. Sugeta et al. (1973) in a study of dialkyl disulfides, which have S–S dihedral angles of 90°, correlated ν_{S-S} with the C–S dihedral angles. Thus ν_{S-S} frequencies of 510, 525, and 540 cm^{-1} are expected when the two trans sites with respect to the sulfur atoms are occupied by two hydrogen atoms, one hydrogen and one carbon, or two carbon atoms, respectively. However, Van Wart and Scheraga (1976) contend that there is no simple relation between ν_{S-S} and rotation about the C–S bonds. Instead, they suggest that after correcting for the effects of substitution on the carbon atom there is a linear correlation between ν_{S-S} and the CS–SC dihedral angle between 0°–60°. For a CS–SC dihedral angle between 70° and 90° these authors place a value of 510 cm^{-1} on ν_{S-S}. Observations by Kitagawa et al. (1979c) on ν_{S-S} from a fragment of a Bence–Jones protein support Sugeta's rather than Scheraga's model. However, Kitagawa et al. make it clear that these conclusions depend on the transferability of X-ray crystallographic data between the same fragments of different Bence–Jones proteins.

Even with the qualifications imposed by differences in interpretation, the S–S stretching region is a very useful monitor of disulfide conformation and concentration, and this spectral region has been of value in studies of peptide hormones and toxins (Maxfield and Scheraga, 1977; Hruby et al., 1978; Tu et al., 1979; Bailey et al., 1979; Ishizaki et al., 1979).

Both cystine (disulfide) and methionine groups in proteins give rise to C–S stretching vibrations in the 600–750-cm^{-1} regions. Hence correlations proposed between ν_{C-S} and conformation within the –C–C–S– moiety in disulfide and methionine are easier to apply when only one group is present. The C–S stretching frequency has been suggested, on the basis of studies on dialkyldisulfides, to lie near 630–670 cm⁻¹, when X (trans to the sulfur atom) in

$$\begin{array}{c} \text{H} \\ | \\ \text{X--C--C--SS} \\ | \\ \text{H} \end{array}$$

is a hydrogen, and 700–745 cm⁻¹ when X is a carbon atom, (Sugeta *et al.*, 1972; Bastian and Martin, 1973). In studies of models of the methionine group, containing ≡C—CH₂—S—CH₃, the observed C–S stretching frequencies were found to vary with the conformation about the C–C and C–S bonds, (Nogami *et al.*, 1975). The correlations suggested by these authors have been used to determine the conformation of C–C–S linkages in neurotoxins (Harada *et al.*, 1976; Takamatsu *et al.*, 1976) and proteins (Chen and Lord, 1976a; Craig and Gaber, 1977). The intensity ratio I_{C-S}/I_{S-S} can be used to follow structural changes of the disulfide group. However, no correlation has been found between this intensity ratio and the C–S–S geometry (Bastian and Martin, 1973; Van Wart *et al.*, 1973).

Cysteine side chains which are not involved in disulfide bridges are usually present in the reduced –SH, or thiol, form. Few protein studies have assigned Raman bands to –SH stretching modes; one exception being the studies by East *et al.* (1978) (discussed in Section III,F,3) of the proteins found in the lens of an eye. For these proteins East and her colleagues clearly identified ν_{S-H} at 2580 cm⁻¹, which is within the range of 2590–2550 cm⁻¹ regarded as "normal" for thiols. The –SH mode has also been identified in the spectra of some virus particles, and it is seen in the Raman spectum of the MS2 virus shown in Fig. 7.8.

2. Tyrosine vibrations

It was suggested by N.-T. Yu *et al.* (1973) that the ratio of the intensities of the tyrosine ring vibrations at 850 and 830 cm⁻¹ (R_{Tyr}) reflects "buried" and "exposed" tyrosines. Subsequently, it was shown that the doublet arises from Fermi resonance between a ring-breathing vibration and the overtone of an out-of-plane ring bending vibration (Siamwiza *et al.*, 1975) and that the intensity ratio is sensitive to the nature of hydrogen bonding or state of ionization of the phenolic hydroxyl group. If the OH group is strongly bound to a negative acceptor, R_{Tyr} is near 0.3. If the OH forms moderately strong hydrogen bonds to H₂O, R_{Tyr} is approximately 1.25, and if the OH participates as a strong hydrogen bond acceptor, R_{Tyr} is close to 2.5. For the –O⁻ form of the phenol side chain R_{Tyr} is near 1.5. The doublet is seen in the spectrum of ribonuclease in Fig. 4.5.

3. Tryptophan modes

A sharp line at 1361 cm⁻¹ has been suggested as an indicator of buried tryptophan residues (Yu, 1974). Thus the presence or absence of a line in this region in the spectrum of proteins containing tryptophan suggests buried or exposed tryptophans. In human carbonic anhydrase B, which is known to have four buried tryptophan residues, the Raman spectrum does indeed show a sharp line at 1363 cm⁻¹ (Craig and Gaber, 1977). A line at 1386 cm⁻¹ from deuterated tryptophan can be used to follow the kinetics of H–D exchange for tryptophan residues in proteins exposed to D_2O. The rate of exposure of the tryptophan to the solvent, attributable to the "opening and closing" of the protein structure, can thereby be calculated (Takesada et al., 1976). Kitagawa et al. (1979c) have proposed that the intensity of a protein Raman feature found at 880 cm⁻¹ can be used as an additional probe of tryptophan environment. Using this band in the spectrum of a Bence–Jones protein, they were able to suggest that one of the protein's two buried tryptophans becomes exposed between pH 4.0 and 3.2 and the other between pH 3.2 and 1.4.

4. Histidine vibrations

The imidazole side chain of histidine often plays an important role in protein function. Unfortunately, intense characteristic imidazole modes have not yet been identified in protein Raman spectra. However, Ashikawa and Itoh (1979) have laid the ground work for such an approach in a careful study of polypeptides containing the L-histidine residue. Moreover, they were able to identify the two tautomeric forms of the neutral imidazole side chain and calculate enthalpy differences between the tautomers in both L-histidine and 4-methylimidazole.

5. Acid groups

Aspartic acid and glutamic acid (Table 4.1) have –COOH side chains. Although the –COOH moiety has a group frequency in the 1700–1750-cm⁻¹ region, features in this part of the spectrum are rarely reported in the Raman spectra of proteins. However, a mode ascribable to the ionized side chains, –COO⁻, is sometimes seen near 1417 cm⁻¹ in the Raman spectra of proteins containing a high percentage of acidic amino acids. The muscle proteins, discussed in Section III,F,4 provide an example of –COO⁻ modes which are an important monitor of protein function.

F. Examples of Proteins Studied by Raman Spectroscopy

The conformational probes, outlined in the previous sections for protein backbone structure and for amino-acid side-chain environment have been put to wide use in protein studies. Quite detailed and specific conformational conclusions can be reached regarding low-molecular-weight proteins, such as toxins and hormones (Tu *et al.*, 1979, and references therein); often conclusions may be drawn at the level of a single side chain in the molecules (e.g., Takamatsu *et al.*, 1976), and there has been considerable activity in this area. Comparative studies on isoenzymes (Twardowski, 1978, 1979) and studies on the effects of protein degradation by ozone (Bieker and Schmidt, 1979) have also been reported. Four different examples are discussed in following sections. These illustrate the use of various regions in the Raman spectrum for conformational analysis and, moreover, emphasize some unique advantages of Raman spectroscopy in protein studies.

1. The thermal denaturation of ribonuclease A

A major advantage of Raman spectroscopy in protein denaturation studies is that several facets of protein conformation can be monitored simultaneously. The study by Chen and Lord (1976) of the thermal denaturation of ribonuclease A provides an elegant example.

The Raman spectra at 32° and 70°C are shown in Fig. 4.5 along with spectral assignments. By monitoring the amide I and III regions Chen and Lord (1976a) were able to show that substantial amounts of helical and pleated-sheet conformations remained at 70°C. Changes in the strength of hydrogen bonding by the tyrosyl residues were followed by monitoring the 830–850-cm^{-1} doublet. As can be seen in Fig. 4.6 changes in the geometry of the disulfide bridges were followed by the frequency and half-width of ν_{S-S} near 510 cm^{-1}. The C–S stretches from cysteines and methionines were used to follow conformational changes in these residues, and Fig. 4.6 also shows the variations with temperature of a C–S peak at 657 cm^{-1}. All probes placed the melting temperature at or near 62°C and indicated a multistep unfolding, rather than an abrupt transition between two states.

2. Raman difference spectroscopy applied to ovalbumin

By using computerized data analysis and reduction it is now possible to subtract the Raman spectra of two similar molecules and detect very small spectral differences. Figure 4.7 shows the Raman spectra of the protein ovalbumin and its close analog S-ovalbumin reported by Kint and Tomimatsu (1979). The spectra appear to be essentially superimposable, but a computer-subtracted difference spectrum, taking away the spectrum of S-ovalbumin from that of ovalbumin, clearly show differences in the amide I region at 1670 cm^{-1} and the amide III region at 1270 and 1235 cm^{-1}. Kint and Tomimatsu ascribe these differences to an increase by 3% of antiparallel β-sheet geometry and a concomitant decrease in α-helical geometry in S-ovalbumin compared to ovalbumin. These researchers

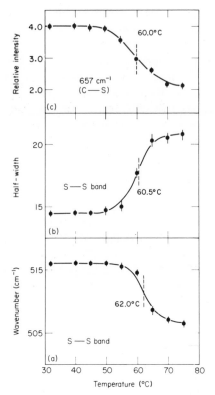

Fig. 4.6. Thermal transition curves of 7% ribonuclease A as monitored by (a) the wavenumber of the band attributable to S–S stretching; (b) the half-width of this band; (c) the relative intensity (peak height) of the band attributable to C–S stretching at 657 cm^{-1}. Reprinted with permission from Chen and Lord, *Biochemistry* **15**, 1889 (1976). Copyright 1976 American Chemical Society.

were able to show that the peaks in the difference spectrum are experimentally significant by illustrating, in a control experiment, the effect of shifting an ovalbumin spectrum by 0.1 cm^{-1} and subtracting it from itself. This example illustrates that very small differences in protein conformation may be detected, and that, in addition, the region of the molecule where the change is occurring may be identified.

3. Proteins in the lens of an eye

Studies of the normal Raman spectra of proteins in the intact lens of an eye demonstrate a unique advantage of the Raman technique. Protein conformation can be monitored *in situ,* and, at the same time, changes in

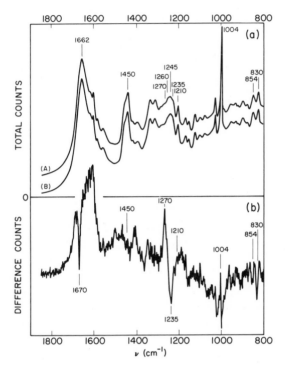

Fig. 4.7. (a) Raman spectra (sum of seven scans) of approximately 7% aqueous solutions, pH 8, of ovalbumin (A) and S-ovalbumin (B); (A) displaced by 10 × 10⁴ units from original (A). (b) Raman difference spectrum of ovalbumin minus S-ovalbumin (obtained from data shown in (a)). The zero is displaced since the ovalbumin spectrum has a higher background than the S-ovalbumin spectrum. (Reproduced from Kint and Tomimatsu, 1979, by permission.)

the Raman spectra from different parts of the lens reflect changes in the relative concentrations of the different proteins found in the lens. The area of the lens being sampled is, of course, simply governed by the diameter of the focused laser beam which is typically 100 μm. Based on Raman measurements, the major protein found in bird and reptile lenses, δ-crystallin, was shown to have predominantly α-helical structure (Kuck *et al.*, 1976; Yu *et al.*, 1977) whereas proteins in bovine and dogfish lenses consist of β-pleated and random-coil structures (Yu and East, 1975). Schachar and Solin (1975) further suggested on the basis of depolarization measurements that in bovine lenses the hydrogen bonded linear CONH groups of the antiparallel β-sheet are preferentially oriented in directions orthogonal to optic axis of the lens.

Yu and co-workers (East *et al.*, 1978) have performed some elegant work on the relative concentration of S–S and –SH groups in the proteins found in the nucleus of the rat lens as a function of age. Comparing the Raman spectrum of a 28-day-old lens with that of a 7-month-old lens, there is a marked decrease in –SH (at 2580 cm^{-1}) which is attributable to cysteine side chains and a corresponding increase in S–S (around 500 cm^{-1}) on aging. These changes are thought to occur in preexisting protein since the constancy of the tyrosine to phenylalanine ratio in the aging lens nucleus is consistent with the idea that nuclear lens proteins retain their original amino acid composition.

The Raman spectra of intact ocular lenses can also be of considerable utility in research on cataract formation. If directly applied to human or animal eyes the Raman probe can monitor precataractous signals or the development of cataracts induced by drugs or radiation. A possible objection to the widespread use of the Raman method is that the laser power required to acquire the spectra, typically 100–300 mW, can cause damage to the eye, especially to the rhodopsin pigment. However, Mathies and Yu (1978) have shown that by using a modified double monochromator and a multichannel optical analyzer (Chapter 3, Section III,C) it is possible to obtain a Raman spectrum from the protein in an eye lens with 1 mW of power for excitation, which is below the threshold for photochemical damage to the eye. Figure 4.8 shows the spectra recorded with 100 mW of 514.5-nm excitation and 29.5 sec data integration time and with 1 mW of 514.5-nm excitation and 655 sec integration time. All the intense Raman lines are attributable to lens protein with major contribution from the aromatic side chains, e.g., the tyrosine doublet at 831 and 855 cm^{-1} is prominent. The S–S bond stretching vibration is expected to appear at 508 cm^{-1}, but is absent in the spectra shown in Fig. 4.8, indicating that S–S bonds do not form in the young bovine lens. Mathies and Yu's work demonstrates that a method is available for acquiring high-quality Raman spectra of an intact lens with very low laser power. This has potential for those engaged in research into cataract formation and, additionally, affords the possibility of clinical application for cataract diagnosis.

4. Conformations of proteins within muscle fibers

Another elegant example of the use of Raman spectroscopy as an *in situ* probe of protein conformation concerns the study of muscle fibers (Asher *et al.*, 1976; Pézolet *et al.*, 1978). Important questions regarding the nature of muscle action center on differences in conformation of protein components in muscle fibers when the muscle is in a relaxed or contracted state. To obtain Raman spectra of intact muscle fiber in both states, Pézolet and co-workers (1980) built the cell shown in Fig. 4.9. The fiber is held in a glass capillary tube, and the relaxed or contracted state is induced by changing the components in the solution bathing the fiber. The

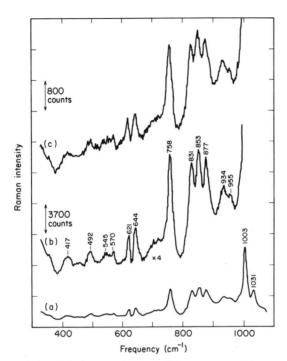

Fig. 4.8. Raman spectrum of low-frequency vibrations of bovine lens protein using multichannel detection. (a) Spectrum taken with 100-mW power, 514.5 nm excitation, and 10.5-cm^{-1} slits. Total integration time 29.5 sec. (b) Fourfold expansion of (a). (c) Spectrum taken with 1-mW power, 514.5-nm excitation, and 14-cm^{-1} slits. Total integration time 655 sec. (Reproduced from Mathies and Yu, 1978, by permission.)

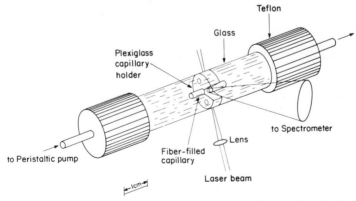

Fig. 4.9. Raman cell for fiber-filled capillaries. The plexiglass capillary holder contains holes to enable the circulation of the equilibrating solution and to let the incident laser beam and the transverse scattered light go through. With this arrangement it is possible to change the composition of the bathing solution without disrupting the sample. The diameter of the laser beam at the sample is approximately 50 μm. (Reproduced from Pézolet *et al.*, 1980, by permission.)

Raman spectra obtained using the device shown in Fig. 4.9 contain features mainly arising from the highly ordered protein lattice within the muscle fiber. The spectral features in the amide I and III regions are quite similar for the relaxed state and for the contracted state induced by addition of ATP and Ca^{2+}. Hence contraction does not bring about a marked change in secondary structure, which is predominantly α-helical as evidenced by the amide I feature at 1650 cm^{-1} and weak intensity in the amide III region. However, addition of ATP and Ca^{2+} does cause a marked intensity decrease of several bands that arise from vibrations of the amino acid side chains of the proteins, especially from the acidic and tryptophan residues. Pézolet and co-workers (1980) ascribe the decrease in intensities of the acidic side chains to strong interactions between carboxylate groups and either positively charged counterions in the mycoplasm or the basic amino acid residues of the proteins. A proposal for the change in intensities of tryptophan modes is that the indole side chains are partially oriented in the muscle fibers and that the preferred orientation changes on contraction.

IV. Resonance Raman Studies of the Chromophores of the Polypeptide Chain Using Excitation below 300 nm

Although, at present, the great majority of RR studies of protein systems concern chromophores absorbing in the visible spectrum, most proteins possess strong absorbance from the aromatic amino acid side chains of tryptophan, tyrosine, and phenylalanine in the 250–280-nm region. These transitions have been of inestimable value to biochemists in assaying for protein concentration or for structural studies using, for example, circular dichroism or fluorescence techniques. Therefore, the march of laser technology into the UV region heralds important new RR applications in the protein field. A technical drawback to UV-excited RR spectra is that luminescent background problems are expected to be more severe than with spectra obtained using visible excitation. This difficulty can be overcome by technological developments, but a more serious objection is the possible loss of selectivity in UV-excited RR spectra. Contributions from many chromophores may make spectral interpretation difficult and systems for study will have to be chosen with care.

An early preresonance Raman experiment on an intact protein showed that side-chain features could be selectively observed. Using 363.8-nm excitation to obtain the preresonance Raman spectrum of lysozyme, Brown *et al.* (1977) reported marked intensity enhancement of certain

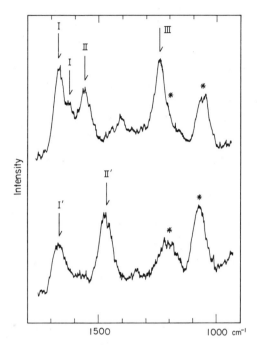

Fig. 4.10. Raman spectra of (a) β-sheet poly-L-lysine film and (b) N-deuterated poly-L-lysine film excited by 257.3-nm irradiation. Asterisks indicate Raman bands attributable to quartz). (Reproduced from Sugawara *et al.*, 1978, by permission.)

indole ring modes of tryptophan. The authors suggested that the enhanced ring modes may be of use in monitoring enzyme reactions involving indole side chains. A feature assigned to a C–S–S–C linkage also appeared to show intensity enhancement, and again this suggests interesting possibilities for following important conformational changes, e.g., those which are thought to occur in antibody molecules in the C–S–S–C bonds. Hirakawa and co-workers (1978) have presented a rigourous RR spectrum, using 257.3-nm excitation, of indole contained in aqueous tryptophan. Under the conditions of rigourous resonance only the mode giving rise to a line at 1623 cm^{-1} was observed. Shimanouchi's group have used the same excitation wavelength, obtained by frequency doubling the 514.5-nm argon ion line, to obtain preresonance Raman spectra of the amide modes in some simple amides (Harada *et al.*, 1975) and in poly-L-lysine and poly-L-glutamic acid (Sugawara *et al.*, 1978). For these molecules, significant intensity enhancement of most of the amide modes shown in Fig. 4.4 was observed, with the additional bonus that the amide II mode, which is very weak or absent in normal Raman protein spectra, was easily identified in the preresonance spectrum. In fact, the amide I, II,

III, I', and II' dominate the 257.3-nm excited spectrum of poly-L-lysine shown in Fig. 4.10. Excitation sources, e.g., excimer lasers, are becoming available in the 190-nm region where a major peptide $\pi \to \pi^*$ transition occurs, and there seems no reason to doubt that rigorous RR spectra of the peptide linkage will soon be reported.

Suggested Review Articles for Further Reading

Frushour, B. G., and Koenig, J. L. (1975). Raman spectroscopy of proteins. *Adv. Infrared Raman Spectrosc.* **1**, 35.

Frushour, B. G., Painter, P. C., and Koenig, J. L. (1976). Vibrational spectra of polypeptides. *J. Macromol. Sci., Rev. Macromol. Chem.* **C15**, 29.

Lord, R. C. (1977). Strategy and tactics in the Raman spectroscopy of biomolecules. *Appl. Spectrosc.* **31**, 187.

Peticolas, W. L. (1979). Low frequency vibrations and the dynamics of proteins and polypeptides. *Methods Enzymol.* **61**, 425.

Van Wart, H. E., and Scheraga, H. A. (1978). Raman and resonance Raman spectroscopy. *Methods Enzymol.* **49**, 67.

Yu, N.-T. (1977). Raman spectroscopy: A conformational probe in biochemistry. *CRC Crit. Rev. Biochem.* **4**, 229. Contains some comments on RR spectra of proteins in addition to the Raman work.

CHAPTER 5

Resonance Raman Studies of Natural, Protein-Bound Chromophores

I. Introduction

Although proteins do not absorb light in the 300–750-nm region, they often have a prosthetic group (e.g., a porphyrin moiety) bound to them which gives rise to an intense visible absorption band and thus makes them available for RR studies. This chapter deals with the RR studies of naturally occurring protein-bound chromophores. Nearly all examples can be grouped into one of three categories: chromophores based on (a) a tetrapyrrole ring (e.g., in the heme proteins), (b) a polyene chain (e.g., retinal of the visual pigments), or (c) protein-bound metal ions, especially those of Fe(III) and Cu(II).

With the exception of metalloproteins, intensity enhancement is usually achieved by excitation into the $\pi \rightarrow \pi^*$ electronic transitions which, for molecules possessing extended π-electron systems, occur in the visible region of the spectrum. Appreciable enhancement also can be obtained from electronic transitions having mixed $\pi \rightarrow \pi^*$, $n \rightarrow \pi^*$ character. The usual definition of metalloproteins excludes the heme proteins. Hence metalloproteins do not possess a porphyrin group; instead, the metal is bound to the protein by amino acid side chains with additional coordination to the peptide backbone sometimes being used. Certain metalloproteins, especially those containing Fe and Cu, have intense ligand-to-metal charge-transfer bands in the visible or near-ultraviolet regions, and it is these transitions which are the source of resonance Raman intensity enhancement. For protein studies, as for any other biochemical system, the RR effect offers the twin advantages of sensitivity and selectivity. The resonance Raman spectrum of a protein-bound chromophore can often be obtained at chromophore concentrations of $10^{-4} M$ or less. At these concentrations the normal Raman spectrum of the protein is undetectable. Thus the RR spectrum provides a selective probe of the chromophore, uncluttered by features from protein or solvent. Moreover, chromophoric sites are often of great functional significance, and the RR spectrum therefore focuses on protein function. Most of the easily obtained, naturally occurring chromophores absorbing in the visible spectral region and not possessing high quantum yields of fluorescence have been studied, and the results are outlined in the discussion that follows. However, it is possible to extend the selectivity and specificity advantages of the RR approach to nonchromophoric systems by a labeling technique (Carey and Schneider, 1978), and this is the subject of Chapter 6.

II. Heme Proteins

The topic receiving the greatest attention in the bio-Raman area during the 1970s and the early 1980s was undoubtably the RR spectroscopy of heme proteins. The popularity of the heme systems stems from their considerable interest to both biochemists and spectroscopists. Reports by Strekas and Spiro (1972a) on cytochrome c and hemoglobin (Strekas and Spiro, 1972b) and an independent report by Brunner *et al.* (1972) on hemoglobin increased interest in the potential of RR spectroscopy for

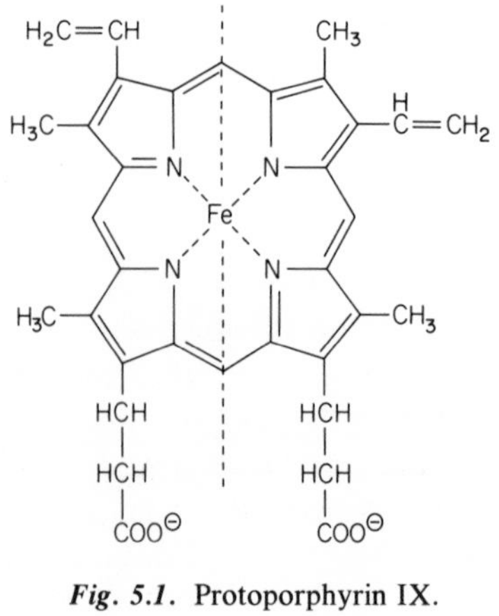

Fig. 5.1. Protoporphyrin IX.

biochemical studies. Additional impetus was added to the spectroscopic aspect by Spiro and Strekas's (1972) report of inverse polarization. This was the first observation of depolarization ratios exceeding 0.75 in vibrational Raman spectroscopy, although they had been predicted some 40 years earlier by Placzek (1934). Since 1972, RR spectroscopy has provided a wealth of new insights into heme structure, and, at the same time, these systems have provided tests for RR theory. In this section the chemistry of the heme groups is outlined, and then their absorption and RR spectral properties (some of which remain unique to the highly symmetric heme chromophores) are discussed. Finally, we show how heme chemistry can be probed by monitoring the marker bands in the 1100–1700-cm^{-1} region or the lower frequency iron–ligand bands.

A. Heme Chemistry

Heme containing proteins play such diverse roles as reversibly binding molecular oxygen for transport (hemoglobin and myoglobin), transferring single electrons in membrane-centered respiratory chains (cytochromes), reducing peroxides (catalases and peroxidases), or acting as the terminal components in multienzyme systems involved in hydroxylations. Usually, the prosthetic group (tightly and sometimes covalently bound to apoprotein) involves an iron atom [Fe(III) or Fe(II)] coordinated to a macrocyclic tetrapyrrole ring (e.g., protoporphyrin IX, Fig. 5.1). The pyrrole nitrogens constitute the four ligands to the iron in equatorial positions, allowing the two axial positions to be occupied by other ligands (Fig. 5.2). Generally, at least one axial position is filled by a ligand from apoprotein. The second axial position may be filled by a ligand from the protein, by H_2O, or by O_2; or by an extraneous ligand, such as CO or CN^-, which may inactivate the protein. In the functional forms of hemoglobin and myoglobin, iron is present as Fe(II) and is not redox active since the release of O_2 in parts of the body where oxygen tension is low is not accompanied by oxidation of Fe(II) to Fe(III). However, both hemoglobin and myoglobin

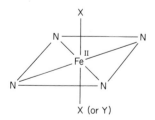

Fig. 5.2. The ligands surrounding the iron atom in a heme group.

can be oxidized to Fe(III) (also known as the hemin) by oxidizing agents such as ferricyanide. The respective products are called *methemoglobin* and *metmyoglobin* and do not function as reversible oxygen carriers. The most intensively studied cytochromes are those found in the mitochondrial membranes in animal cells, where they act as one-electron carriers between various iron–sulfur flavoprotein dehydrogenases and molecular oxygen. Cytochromes *a, b,* and *c* have different macrocycles owing to differing side-chain substitutions in the basic structure shown in Fig. 5.1. The varying protein environments about each macrocycle, together with differences in the macrocycles themselves, account for variations in redox potentials found among the cytochromes. Cytochromes have recently been isolated which are functionally distinct from the classical mitochondrial cytochromes, and examples of these nonmitochondrial systems are cytochrome b_5 which is involved in the desaturation of stearyl-CoA and the family of P_{450}-type cytochromes which activate O_2 for hydroxylation reactions.

The oxidation and spin states (and consequently the size) of the iron atom bound to the tetrapyrrole ring can play an important role in the interpretation of RR spectra of hemes, and these characteristics must be dealt with before consideration is given to the spectroscopic properties. Two oxidation states of iron, Fe(II) and Fe(III), are stable in aqueous solutions and these are the major redox forms in iron proteins. Both oxidation states can have high or low spin states. Fe(II) has six valence electrons in its 3d orbitals while Fe(III) has five valence electrons. There are five 3d orbitals, and in a nonperturbing environment (e.g., in the gas phase) these have the same energy and are thus degenerate. Since an amount of energy E_p must be expended to persuade electrons to pair up, the occupancy of the degenerate 3d orbitals for Fe(II) and Fe(III) will be

Fe(III)   ↑   ↑   ↑   ↑   ↑

Fe(II)   ↑↓   ↑   ↑   ↑   ↑   degenerate 3d orbitals

However, in the biological context, iron atoms are surrounded by ligands, and the ligands remove the degeneracy of the 3d orbitals. For the octahedral ligand field (Cotton and Wilkinson, 1980) of the type found in hemes, the orbitals are split into two discrete groups of three low-lying and two higher energy orbitals separated by a ligand-field energy Δ.

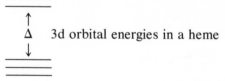

↑
Δ   3d orbital energies in a heme
↓

The way in which the 3d electrons fill the orbitals depends on the relative magnitudes of Δ and the electron pairing energy E_p. When Δ is small relative to the unfavorable pairing energy E_p, then all five 3d orbitals will be populated; this gives *high-spin iron*. High-spin Fe(III) has five unpaired electrons; high-spin Fe(II) has four unpaired electrons

High-spin Fe
$(\Delta < E_p)$

     Fe(III)   Fe(II)

When the magnitude of the field-splitting energy is significantly greater than the spin-pairing energy, then the 3d electrons distribute themselves among the three low-lying orbitals, resulting in *low-spin* iron

Low-spin Fe
$(\Delta > E_p)$

     Fe(III)   Fe(II)

In the low-spin state, Fe(II) has no unpaired electrons, while Fe(III) has one. Since the higher energy orbitals are unoccupied in low-spin Fe [Fe(II) or Fe(III)], the low-spin iron atom is smaller than the high-spin iron. This change of size of Fe with change in spin state plays an important role in establishing iron–macrocycle geometries. The central "hole" in the tetrapyrrole macrocycle is 2.02 Å in diameter; thus low-spin Fe(II) or Fe(III) (~1.91 Å) can fit within the cavity and lie in the plane of the tetrapyrrole. In contrast, high-spin iron is ~2.06 Å in diameter and consequently must sit above the cavity (see Fig. 5.3). High-spin Fe(III) lies about 0.3 Å above the plane, and high-spin Fe(II) lies about 0.7 Å above

Fig. 5.3. (a) Low-spin Fe. (b) High-spin Fe.

the plane. Several heme systems seem to be balanced at the transitions between high-spin and low-spin states, and the axial ligands can determine the equilibrium position by varying the ligand-field splitting, Δ. For example, when hemoglobin binds oxygen, the oxygen coordination induces a shift from high-spin Fe(II) to low-spin Fe(II) · O_2 allowing the iron atom to move closer to the plane of the macrocycle.

B. Heme Absorption and RR Spectroscopy

The structure of a RR spectrum attributable to a heme protein depends on the absorption band chosen to generate the RR effect. It is therefore necessary to outline the general characteristics of heme absorption spectra. The visible and near-ultraviolet absorption spectra of hemes and metalloporphyrins are dominated by two $\pi \rightarrow \pi^*$ electronic transitions. Both transitions are polarized in the plane of the heme and are of the same symmetry. The $\pi \rightarrow \pi^*$ transitions are subject to strong configuration interaction, with the result that the transition dipoles are additive for the higher energy transition and largely cancel for the lower energy one. A typical spectrum resulting from the $\pi \rightarrow \pi^*$ transitions is shown in Fig. 5.4 for cytochrome c in the reduced form. The higher energy transition is

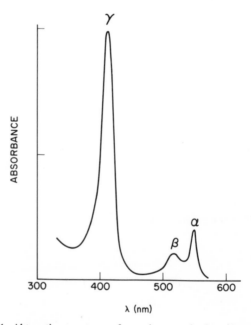

Fig. 5.4. Absorption spectrum of cytochrome c in the reduced form.

assigned to the intense ($\epsilon \sim 10^5 \, M^{-1} \, cm^{-1}$) absorption band, called the Soret or γ band, near 400 nm (Fig. 5.4). The lower energy transition is assigned to the α band in Fig. 5.4; as expected on the basis of near cancellation of transition dipoles, the α band is an order of magnitude less intense than the Soret band. However, the lower energy transition can "borrow" some of the intensity of the higher energy transition through vibrational interactions. These produce a vibronic side band, called the β band, which occurs at an energy $\sim$1300 cm^{-1} higher than the α band. The α band is caused by a pure electronic transition in which no change in vibrational quantum number occurs (i.e., the $v'' = 0$ to $v' = 0$ transition in Fig. 2.4). The β band is attributable to the envelope of all the active *vibronic* transitions in which the vibrational quantum numbers of the vibronically active modes increase from 0 to 1 (i.e., the $v'' = 0$ to $v' = 1$ jump in Fig. 2.4). Additionally, heme and metalloporphyrin charge-transfer transitions can occur in the electronic spectrum. The presence of charge-transfer electronic transitions is reflected by extra absorption bands or by a broadening of the α, β, or Soret bands. The magnitude of the charge-transfer contribution to the absorption spectrum is related to the energy difference between the metal d and porphyrin π orbitals and the identity of the fifth and sixth axial ligands.

1. Excitation into the Soret region

Excitation into the Soret region is expected to produce intense RR peaks from only totally symmetric modes. Thus, according to Chapter 2, Section VII, the depolarization ratios ρ should be between 0 and 3/4, and symmetry considerations indicate that ρ should be 1/8 for a D_{4h} macrocycle (Sonnich Mortensen, 1969; Felton and Yu, 1978). However, the measured values of ρ are seldom exactly equal to 1/8 (although, for excitation into the Soret band, they are less than 3/4), and several reasons, such as the breakdown of local symmetry, have been advanced to explain this (Remba *et al.*, 1979). The mechanism for resonance enhancement in the Soret region is thought to be attributable to the Albrecht A type (Chapter 2, Section VIII) with involvement of a single excited electronic state and great intensity enhancement of those modes which change the geometry of the porphyrin macrocycle in a way that mimics the changes occurring on going from the ground to the excited electronic state. Thus Remba *et al.* (1979), using Soret-band laser excitation for some low-spin forms of chloroperoxidase, observed enormous selective enhancement of features at 1360 and 674 cm^{-1} which they ascribed to normal modes of expansion of the porphyrin moiety on electronic excitation. Some support for an A-type mechanism also comes from excitation profiles. For example the preresonance excitation profile of the 1373-cm^{-1} band from ferrihemoglobin (for which $\rho = 1/8$ showed a frequency dependence (Strekas *et al.*, 1973) in close agreement with that predicted by the A term (Chapter 2, Section VIII).

2. Excitation into the $\alpha-\beta$ region

Excitation into the $\alpha-\beta$ region produces intense RR bands which are either depolarized ($\rho = 0.75$) or *anomalously polarized* ($\rho > 0.75$), and the source of intensity enhancement is thought to be vibronic and involve an Albrecht B-type mechanism. Accordingly, those normal modes which mix the α and Soret transitions (which are just those responsible for the β band) are expected to undergo resonance enhancement. By exciting into

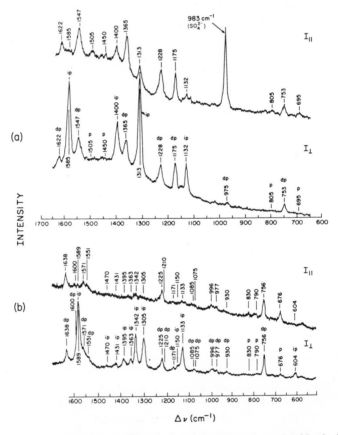

Fig. 5.5. Resonance Raman spectra of (a) ferrocytochrome c, 0.5 mM, obtained with 514.5-nm excitation and (b) oxyhemoglobin, 0.5 mM, obtained with 568.2-nm excitation. [Adapted from Spiro and Strekas (1972).]

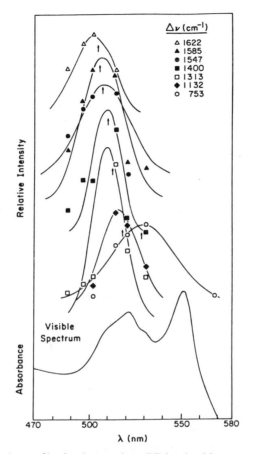

Fig. 5.6. Excitation profiles for the prominent RR bands of ferrocytochrome c and the electronic absorption spectrum, both on a logarithmic scale. The Raman band intensities were measured relative to the 983-cm^{-1} sulfate band as an internal standard. [From Spiro and Strekas (1972).]

or near the β band of ferrocytochrome c and oxyhemoglobin, Spiro and Strekas (1972) obtained the RR spectra shown in Fig. 5.5. The spectra are dominated by peaks in the 1100–1650-cm^{-1} region, because these peaks are due to the in-plane stretching of C–C and C–N bonds and the bending of C–H bonds which are effective in vibronically mixing the in-plane α and Soret transitions.

The excitation profiles through the β band of ferrocytochrome c (Spiro and Strekas, 1972) and the α band of oxyhemoglobin (Strekas and Spiro, 1973) are shown in Figs. 5.6 and 5.7, respectively. For oxyhemoglobin the

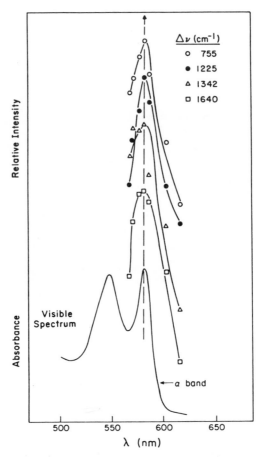

Fig. 5.7. Excitation profiles for the prominent RR bands of oxyhemoglobin and the electronic absorption spectrum. [From Strekas and Spiro (1973).]

profiles peak at the α absorption maximum (the electronically pure 0–0 transition), while the profiles through the β band of ferrocytochrome c shift to lower frequency with a lowering in frequency of the mode involved. The experimental profile maxima within the β band are in reasonable agreement with predictions based on the maxima occurring at the 0–1 (i.e., $v'' = 0$ to $v' = 1$ in Fig. 2.4) vibronic transition. Thus the separation between the α absorption maximum (0–0) and the 0–1 maxima, revealed by the excitation profiles, should yield the vibrational frequency in the excited electronic state. For a large chromophore, such as a porphyrin, many normal modes will have similar excited and ground state frequencies. The observation of maxima at both the 0–0 and 0–1 frequencies of the same heme protein was not possible at the time the work shown in Figs. 5.6 and 5.7 was conducted because of the restricted laser wavelengths available. However, maxima at both the 0–0 and 0–1 frequencies have been observed subsequently [e.g., Mn(III)–etioporphyrin I (Shelnutt *et al.*, 1976)]. Excitation profiles in the α–β-band region have also been used to estimate excited-state lifetimes in cytochromes c and b_5. By comparing excitation profiles with absorption spectra (Penner and Siebrand, 1976) and by comparing excitation-profile bandwidths, Friedman *et al.* (1977) concluded that the excited-state lifetime in cytochrome c is longer ($\sim 45 \times 10^{-15}$ sec) than in cytochrome b_5 ($\sim 30 \times 10^{-15}$ sec).

We can now elaborate on the statement made at the beginning of this section that excitation into the α–β region results in intense depolarized (dp) or anomalously polarized (ap) Raman bands. This is shown for ferrocytochrome c and oxyhemoglobin in Fig. 5.5. Only very weak polarized (p) bands are seen in the 514.5-nm excited spectra of Fig. 5.5. The values of the depolarization ratios are governed by symmetry considerations. Only an outline of the symmetry properties is given here; for a more detailed description of group theory and the role of symmetry in spectroscopy the reader is referred to the monograph by Cotton (1971). The vibronic scattering mechanism observed in the α–β region requires that the symmetries of the active vibrations be contained in the direct product of the group theory representations, Γ, of the Soret and α transitions (since these are the transitions undergoing vibronic mixing). The α and Soret transitions are both of E_u symmetry; thus

$$\Gamma = \Gamma_{E_u} \times \Gamma_{E_u} = \Gamma_{A_{1g}} + \Gamma_{A_{2g}} + \Gamma_{B_{1g}} + \Gamma_{B_{2g}}$$

McClain (1971) has tabulated the scattering tensors and hence the expected polarizations for modes with each of these symmetry classes for the D_{4h} point group of the tetrapyrrole macrocycle. The A_{1g} modes are not vibronically active and thus are not intensity enhanced. The depolarized modes in Fig. 5.5 are thought to originate from the B_{1g} and B_{2g} terms which, on the basis of group theoretical arguments, should be depolarized. The scattering tensor (see Chapter 2, Section III) for the A_{2g} modes is most unusual in that it is antisymmetric since for example $\alpha_{xz} = -\alpha_{zx}$. The A_{2g} modes are forbidden in normal Raman scattering and only become allowed under resonance conditions. Detailed considerations of the antisymmetric tensor and the general relationship for the depolarization ratio (Chapter 2, Section VII) show that the A_{2g} modes give rise to inversely polarized bands with $\rho = \infty$. For the inversely polarized bands the plane of polarization of incident light is rotated by 90° on scattering (Fig. 2.13).

Some of the bands marked "ip" (for inversely polarized) in Fig. 5.5 have $0.75 < \rho < \infty$ rather than $\rho = \infty$. The features with $\rho > 0.75$ are usually designated anomalously polarized, reserving the term inversely polarized for those bands for which no parallel component can be detected. For the anomalously polarized bands there is a residual scattering intensity parallel to the incident polarization. This could be attributable to the effective symmetry of the macrocycle being less than D_{4h} or the accidental coincidence of two modes of different symmetry at the same frequency. Additionally, the depolarization ratios of some bands in cytochrome c and other heme proteins show dispersion behavior since they have values which depend on the excitation wavelength (Collins et al., 1976). The phenomenon of dispersion and the failure of bands to obey symmetry rules strictly mean that caution must be exercised in using depolarization data, obtained at a single excitation frequency, to assign bands to symmetry species.

C. Porphyrin Modes As Marker Bands for the Iron Atom

It is apparent from Fig. 5.5 that the RR spectra of heme proteins contain a rich assortment of porphyrin vibrational modes. It was soon realized that certain of these are sensitive to chemical alteration of the central iron atom and could be used as potential structure monitors, (Yamamoto et al., 1973; Spiro and Strekas, 1974). For example, a change of oxidation state of the iron from Fe(III) to Fe(II) is accompanied by a lowering of some porphyrin skeletal vibrational frequencies. Moreover, a change from low-spin to high-spin iron at constant oxidation state is accompanied by an even more marked decrease of other vibrational frequencies. The classification of oxidation and spin-state markers given originally by Spiro and Strekas (1974) and expanded by Spiro and Burke (1976) is given in Table 5.1. The oxyhemoglobin and ferrocytochrome c bands cited in Table 5.1 can be identified in Fig. 5.5

The classification outlined in Table 5.1 has found wide utility in monitoring the chemistry in a number of heme systems. However, there remained anomalies in the categorization, and considerable effort has been devoted to formulating a consistent model in which every heme's RR spectrum in the $1100-1700$-cm^{-1} marker-band region can be accounted for in terms of the properties of the iron. The crucial point is that the iron is being monitored only indirectly, i.e., via the effect its chemical properties have on the porphyrin modes. Therefore it is necessary to understand how the geometry of the porphyrin relates to the properties of the iron. Having a detailed knowledge of the porphyrin geometry and its normal vibrational modes, it is then possible to correlate changes in the properties of the iron with shifts in the marker bands.

Table 5.1

Comparison of Structure and Oxidation-State Marker Bands (cm⁻¹) for
Iron Mesoporphyrin (IX) Dimethyl Ester with Heme Proteins[a]

	I[b]	II[b]	III[c]	IV[d]	V[e]
Low-Spin Fe(III)					
$(\text{Im})_2\text{Fe}^{\text{III}}(\text{MP})$	1375	1505	1572	1584	1640
$(\text{CN}^-)\text{Fe}^{\text{III}}\text{Hb}$	1374	1508	1564	1588	1642
$\text{Fe}^{\text{III}}\text{Cytochrome } c$	1374	1502	1562	1582	1636
$(\text{CN}^-)\text{Fe}^{\text{III}}\text{HRP}$	1375	1497	1562	1590	1642
Low-Spin Fe(II)					
$(\text{pip})_2\text{Fe}^{\text{II}}(\text{MP})$	1358	1490	1537	1583	1620
$(\text{CN}^-)\text{Fe}^{\text{II}}\text{HRP}$	1362	1498	1545	1587	1620
$\text{Fe}^{\text{II}}\text{cytochrome } c$	1362	1493	1548	1584	1620
$(\text{CO})(\text{py})\text{Fe}^{\text{II}}(\text{MP})$	1371	1497	1567	1588	1630
$(\text{CO})\text{Fe}^{\text{II}}\text{Hb}$	1372			1584	1631
$(\text{O}_2)\text{Fe}^{\text{II}}\text{Hb}$	1377	1506	1564	1586	1640
High-Spin Fe(II)					
$(2\text{-MeIm})\text{Fe}^{\text{II}}(\text{MP})$	1359	1472	e	1558	1606
$\text{Fe}^{\text{II}}\text{Hb}$	1358	1473	e	1552	1607
$\text{Fe}^{\text{II}}\text{HRP}$	1358	1472	e	1553	1605
High-Spin Fe(III)					
$(\text{X}^-)\text{Fe}^{\text{III}}(\text{MP})$	1374	1495	1565	1572	1632
$(\text{H}_2\text{O})\text{Fe}^{\text{III}}\text{HRP}$	1375	1500	1550	1575	1630
$(\text{F}^-)\text{Fe}^{\text{III}}\text{Hb}$	1373	1482	1565	1555	1608
Intermediate-Spin Fe(II)					
$\text{Fe}^{\text{II}}(\text{MP})$	1373	1506	1570	1589	1642

[a] Reprinted with permission from T. G. Spiro and J. M. Burke, *J. Am. Chem. Soc.* **98**, 5482 (1976). Copyright 1976 American Chemical Society.

[b] Polarized.

[c] Depolarized.

[d] Anomalously polarized.

[e] Obscured by band IV.

1. The oxidation-state marker bands

The so-called band I, which occurs between 1358 and 1377 cm⁻¹ (Table 5.1), is the most widely used oxidation-state marker. The positions of band I and the other markers in this family are thought to reflect the electron population in the porphyrin π^* orbitals, with an increasing population weakening the porphyrin bonds and bringing about a decrease in vibrational frequency (Kitagawa *et al.*, 1975, 1978; Spiro and Burke, 1976). The electron population in the π^* orbitals is increased by back donation of electrons from the iron atom's $d\pi$ orbitals; since back donation is greater for Fe(II) than for Fe(III), the oxidation-state marker bands for Fe(II) hemes are lower in frequency compared to those of the Fe(III) analogs. In this model, any effect, in addition to a change in oxidation state, which perturbs the distribution of electrons in the porphyrin π^* orbitals can bring about a change in oxidation-state marker frequencies. In this regard, large perturbations to "oxidation-state marker" frequencies occur when an axial ligand possesses π orbitals capable of interacting with the porphyrin orbitals via the iron $d\pi$ electrons. An axial π electron donor will force additional iron $d\pi$ electrons into the porphyrin π^* orbitals and lower the oxidation-state band markers to "atypical" values. Thus the anomalous positions of the peaks in the RR spectrum of cytochrome P_{450} were explained by Ozaki et al. (1976, 1978) and Champion et al. (1978) by the presence of a strongly electron-donating axial ligand thought to be the thiolate ion of a cysteine residue.

2. The spin-state marker bands

A prime example of a spin-state marker band is band IV which, as can be seen in Table 5.1, occurs between 1552–1590 cm^{-1}. Band IV is easily recognized by its high intensity and anomalous polarization. It is immediately identified near 1585 cm^{-1} in the spectra of cytochrome c and oxyhemoglobin in Fig. 5.5. Overall, the spin-state markers have been the subject of a controversy concerning the nature of the structural change in the iron–porphyrin core responsible for bringing about shifts in marker frequencies. A sound correlation between porphyrin structure and band frequencies permits one to monitor the chemistry of the iron–porphyrin core in heme proteins and, in addition, to reach conclusions on such important topics as the effect of the protein matrix on the porphyrin structure. Hence considerable effort has gone into attempts at unambiguously establishing the structure-frequency shift relationship.

Spiro's group originally interpreted the frequency decreases observed for high-spin hemes in terms of increased doming of the ring (Spiro and Burke, 1976, and references therein). As indicated in Section II,A the larger size of the high-spin iron results in out-of-plane displacement of the atom toward an axial ligand. In order to maintain overlap with the iron orbitals, the pyrrole rings are expected to tilt as shown in Fig. 5.8; however, this tilting results in poorer π conjugation through the porphyrin methine bridges and consequently lowers the observed frequency. A different interpretation based on expansion of the porphyrin core (i.e., a lengthening of the distance Ct–N in Fig. 5.8) was advanced by Spaulding et al., (1975; Yu, 1977; Lanir et al., 1979). Spaulding and co-workers found an inverse correlation between the frequency of the anomalously polarized band in the 1580-cm^{-1} region (corresponding to the bands at 1585 and 1589 cm^{-1} in Fig. 5.5) and Ct–N in a number of porphyrin

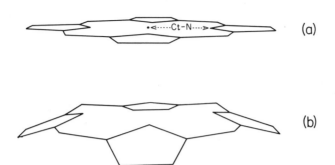

Fig. 5.8. (a) Planar and (b) domed tetrapyrrole skeleton.

derivatives. This correlation, which more recently has been extended to bands V (Lanir *et al.*, 1979; Spiro *et al.*, 1979) and II (Spiro *et al.*, 1979), is shown in Fig. 5.9. Normal coordinate analysis (Abe *et al.*, 1978) provides a reasonable rationale for the change in frequency of the 1580-cm^{-1} A_{2g} band with core size (Spaulding *et al.*, 1975).

Resonance Raman studies of porphyrin model compounds (Mendelsohn *et al.*, 1975b; Spaulding *et al.*, 1975; Spiro and Burke, 1976; Felton and Yu, 1978) measured in the absence of protein have gone hand-in-hand with the investigation of heme proteins. The model porphyrin studies are not described in detail here, except to point out that they provide the crucial structure–spectra correlation because very accurate structural data, based on X-ray crystallography, can be gained from model porphy-

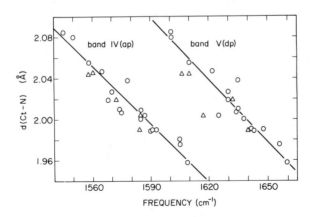

Fig. 5.9. Correlation between the frequencies of bands IV (anomously polarized, ap) and V (depolarized, dp) with core size as defined by the Ct–N distance shown in Fig. 5.8. Reprinted with permission from A. Lanir *et al.*, *Biochemistry* **18**, 1656 (1979). Copyright 1979 American Chemical Society.

rins. In contrast, the structural data on those heme-proteins which have
been characterized by X-ray crystallography are usually less precise. In
fact, the crystal structures of two porphyrins, taken with additional data,
appear to resolve the debate over the "doming" and "core-expansion"
models. The crystal structure determinations of high-spin bisaquo and
bis(tetramethylene sulfoxide) complexes of Fe(III)-tetraphenylporphyrin
showed the porphyrins to be planar with expanded cores. The Raman data
of closely analogous compounds showed that the spin-marker bands for a
planar, expanded-core porphyrin did occur at appreciably lower frequencies compared to out-of-plane five coordinate derivatives with the same
oxidation and spin states (Spiro *et al.,* 1979). Hence Spiro reevaluated the
doming hypothesis, and in his 1979 paper with Stong and Stein supported
core expansion as the dominant cause of a drop in spin-marker frequencies, although doming may have a secondary effect on the band positions.
The coordination number around the iron atom also appears to be important since spin-marker frequencies are appreciably lower for six-
coordinate compared to five-coordinate high-spin hemes, both for Fe(III)
and Fe(II) (Spiro, 1980).

Earlier in this section we emphasized that care must be exercised in
drawing conclusions regarding iron-spin and oxidation-state parameters in
terms of the frequencies of porphyrin modes since the porphyrin modes
provide an indirect monitor. Ironically, it is becoming evident that the
porphyrin modes may be most sensitive to an effect one step removed
from the iron itself—namely, the axial ligands. The presence of a π-donor
axial ligand strongly perturbs the oxidation-state markers, and single or
double axial ligation (giving five- or six-coordinate hemes) controls the
spin markers via core expansion. Any interpretation of marker frequencies must therefore take the number and properties of the axial ligands
into account in addition to iron oxidation and spin state. The marker
bands have already found wide use in characterizing heme proteins, and
as our knowledge of the structural origin of the band shifts improves, the
conclusions on the heme stereochemistry will become increasingly detailed. The next section discusses some recent studies which have utilized
the marker bands to great effect.

D. The Use of Marker Bands in Studying Heme–Protein Interactions and Heme Photolysis

Each molecule of hemoglobin has four peptide chains and four hemes,
each of which can bind an oxygen molecule. The binding of the first
oxygen molecule to one heme group enhances the binding of successive

oxygen molecules to the three remaining hemes. Unliganded mammalian hemoglobins are in the low-oxygen-affinity form, denoted the T state. The binding of oxygen or other ligands is normally followed by a transition to a different, high-oxygen-affinity conformation, denoted the R state, and the cooperative ligand-binding properties of hemoglobin are a consequence of the reversible transition between these two forms. Since the four heme groups of hemoglobin are spaced far apart and cannot interact directly, the means by which the binding of one oxygen molecule changes the affinity for oxygen of a distant heme poses an interesting problem. One of the related questions addressed by RR studies is whether the switch from the T to R form brings about structural changes in the porphyrin core, which, in turn, may be detected by frequency changes in the porphyrin modes. In the 1970s several groups looked for differences in the heme RR spectra of the R and T quaternary states (Scholler and Hoffman, 1979; Shelnutt *et al.*, 1979, and references therein) and were in agreement in finding that no peaks moved by more than 2 cm^{-1} following the R to T transition. However, Shelnutt and co-workers (1979), using a sensitive difference technique that requires the partitioned, rotating cell described by Kiefer (1973), were able to reliably detect changes in peak positions of less than 2 cm^{-1}. For the anomalously polarized band at 1556 cm^{-1} the difference in frequency between the T and R structure was found to be less than ± 1 cm^{-1}. Using the correlation of Spaulding *et al.* (1975) (see also Fig. 5.9) of this frequency with core size, it is apparent that the pyrrole nitrogen-to-center distance does not change by more than 0.002 Å. From this, Shelnutt *et al.* (1979) concluded that any movement of the heme iron from the plane of the pyrrole nitrogens, associated with the R–T transition in deoxyhemoglobin, is not caused by alterations in core size. Nevertheless, Shelnutt and co-workers detected frequencies differences of up to 2 cm^{-1} in all of those modes known to be sensitive to electron density in the antibonding π^* orbitals (Section II,C,1). In particular, the 1357-cm^{-1} marker band shifted by as much as -1.3 cm^{-1} on changing the quaternary structure of hemoglobin from T to R. The mechanism by which protein–porphyrin interactions change the electron population in the π^* orbitals is at this time speculative, but Shelnutt and co-workers (1979) favor a charge-transfer interaction involving the aromatic side chain of certain amino acids and the tetrapyrrole ring.

A second active area of heme chemistry where the marker bands have been carefully studied is the photolysis of heme complexes. In particular, several laboratories have investigated the photodissociation of car-boxyhemoglobin (HbCO). The HbCO complex is dissociated with high-quantum efficiency by visible light, and the transient species so formed have been characterized by absorption spectroscopy on time scales as short as picoseconds. Several groups, using pulsed lasers and optical multichannel detectors, have independently reported the RR spectrum of a transient species of hemoglobin occurring within nanoseconds of the photodissociation of HbCO (Dallinger *et al.*, 1978; Lyons *et al.*, 1978; Woodruff and Farquharson, 1978). The spectrum generated using a single 7-nsec pulse both to photolyse and examine the sample is shown in Fig. 5.10. The RR spectrum generated within the pulse width is essentially that of deoxyhemoglobin, although small, but measurable, frequency lowerings with respect to deoxyhemoglobin have been observed (Lyons *et al.*, 1978). The research has been extended to shorter time frames using 30-

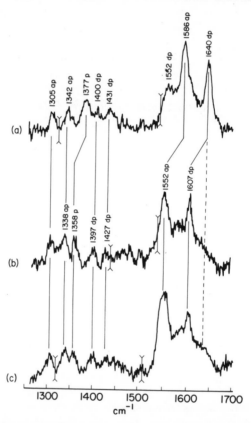

Fig. 5.10. Time-resolved RR spectra of (a) oxyhemoglobin (b) deoxyhemoglobin, and (c) photodissociated carboxyhemoglobin. The spectra were obtained on a multichannel detector by accumulating a series of 7-nsec pulses. The double-ended arrows denote the points where adjacent vidicon frames were joined to make complete spectra. Reproduced from Woodruff and Farquharson (1978). Copyright 1978 by the American Association for the Advancement of Science.

psec pulses (Terner *et al.,* 1980) and the frequency differences are still observed. For example, peaks occur at 1542 (depolarized), 1552 (anomalously polarized), and 1603 cm^{-1} (depolarized) in the RR spectrum of the picosecond transient of HbCO compared to 1549, 1558, and 1607 cm^{-1} in deoxyhemoglobin. The 1542- and 1603-cm^{-1} positions suggest an expanded porphyrin core, and Terner *et al.* (1980) ascribe the low frequencies of the transient to a high-spin Fe(II) species with the iron in the plane of the porphyrin ring.

The picosecond RR spectra of the transient resulting from photolysis of HbCO show that electronic rearrangement in the porphyrin core to give the deoxyhemoglobin-like species must occur in less than 20 psec. Coppey *et al.* (1980) were able to reach a similar conclusion on the deligated transient formed from oxyhemoglobin using 30-psec laser pulses. In a separate study, Kitagawa and Nagai (1979) showed that a reversible photoreduction of hemoglobin took place in the T state but not the R state. In each case, the positions of the marker bands in the 1200–1700-cm^{-1} region were used as the basis of interpretation. By excitation with light near the Soret region, photoreduction of the heme of cytochrome oxidase has also been observed (Salmeen *et al.,* 1978; Adar and Erecínska, 1979). However, the electron donors involved are thought to be different for cytochrome oxidase and hemoglobin. An exogenous flavin is thought to be the donor in cytochrome oxidase while an endogenous species, such as an axial ligand of the iron may be responsible in hemoglobin.

E. Iron–Ligand and Iron–Nitrogen Modes

The most direct way of probing the bonding about the iron atom in a heme group is to monitor the modes attributable to the iron–ligand vibrations which occur below 700 cm^{-1}. Hence the low-frequency RR spectrum holds considerable promise for quantifying bond strain and changes in length for the axial ligand linkages. However, the literature on this frequency region has been slower to develop compared to that on the 1100–1700-cm^{-1} region since, for reasons such as the availably of suitable laser lines, good quality spectra were initially difficult to obtain. A further difficulty lies in assignment, e.g., there is not yet agreement in assigning porphyrin deformation modes, iron–nitrogen (pyrrole) stretches, and iron-nitrogen (axial imidazole) stretches in myoglobin (Desbois *et al.,* 1979; Kincaid *et al.,* 1979a,b). The easiest modes to assign are generally those attributable to iron-axial ligand stretches since large frequency shifts occur when isotopically substituted ligands are used. In this regard Brunner (1974) achieved an early success by identifying the Fe–O$_2$ stretching mode in oxyhemoglobin. As the spectra in Fig. 5.11 show, the

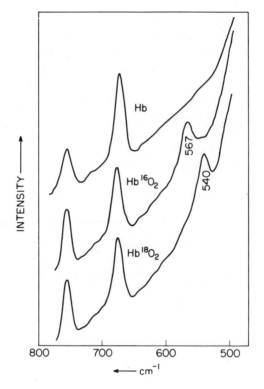

Fig. 5.11. RR spectra, obtained with 488-nm excitation, of $3 \times 10^{-5} M$ hemoglobin (Hb), $Hb^{16}O_2$ and $Hb^{18}O_2$. Note shift in 567-cm^{-1} peak. [From Brunner (1974).]

line at 567 cm^{-1} in Hb$-^{16}O_2$ shifts to 540 cm^{-1} when hemoglobin is saturated with $^{18}O_2$. Brunner also stated that the appearance of a single Fe$-O_2$ stretch in the RR spectrum favored the end on structure **I** rather than the side on structure **II** and Duff *et al.* (1979) were able to confirm Brunner's

$$\begin{array}{ccc}
O^{\nearrow O} & \qquad & O{-}{-}O \\
\mid & & \diagdown\diagup \\
Fe & & {-}Fe{-} \\
\\
\textbf{I} & & \textbf{II}
\end{array}$$

prediction by using $^{18}O{-}^{16}O$. The RR spectrum of the Hb$-^{16}O^{18}O$ complex showed two features at 567 and 540 cm^{-1} which demonstrated that the two ends of the bound dioxygen are located in different environments, and provided strong support for structure **I**. This same conclusion was reached on the basis of the frequency and intensity, of the O–O stretch observed in the infrared spectrum of hemoglobin in erythrocytes (Barlow *et al.,* 1973). The predominant mechanism for Raman intensity enhancement of the iron-axial stretching modes appears to be coupling of the iron-axial vibrations to ligand-to-iron charge transfer bands, although the latter may be mixed with $\pi \rightarrow \pi^*$ transitions (Chottard and Mansuy, 1977). By exciting into charge transfer bands, Asher and co-workers (Asher *et al.,* 1977c; Asher and Schuster, 1979) obtained detailed data on methemoglobin derivatives with OH⁻, F⁻, and N₃⁻ axial ligands. The OH⁻ and F⁻ complexes have different iron-axial ligand frequencies in hemoglobin and myoglobin [both in the Fe(II) states], and Asher interpreted the frequency differences in terms of an increase in the out-of-plane iron distance for the high-spin species of myoglobin (iron-axial ligand modes were not observed for the low-spin species).

A wide range of isotopically substituted analogs were used by Desbois *et al.* (1979) to investigate the low-frequency region of oxidized and reduced horse myoglobin. These investigators used $H_2^{16}O$–$H_2^{18}O$, $^{16}OH^-$–$^{18}OH^-$, $^{14}N_3^-$–$^{15}N_3^-$ and [$^{14}N_2$]-imidazole–[$^{15}N_2$]-imidazole, together with substitution of the central iron; ^{54}Fe was used to replace ^{56}Fe. Ten features, located between 180 and 700 cm⁻¹, which are common to every myoglobin derivative were categorized by Desbois and co-workers; these are listed in Table 5.2. The assignments are tentative and in some cases controversial (Asher and Schuster, 1979; Kincaid *et al.,* 1979a,b); however, Desbois and co-workers concluded that at least five of the low-frequency bands of the RR spectra obtained by excitation with blue irradiation involve iron–ligand modes and can be used directly to monitor coordination number, oxidation state, and spin state of the iron atom.

III. Vitamin B₁₂ (Cobalamin)

Much of the current interest in the chemistry and biochemistry of vitamin B₁₂ and its derivatives stems from the efficacy of the vitamin in controlling pernicious anemia. Structurally, vitamin B₁₂ contains a cobalt atom coordinated to four nitrogen atoms of a corrin ring (Fig. 5.12). One axial position about the Co atom is occupied by benzimidazole, bound covalently to the ring, although this may be removed chemically in, for example, the cobinamides. The sixth coordination position about the cobalt can be occupied by any of a number of ligands.

The corrins have spectral properties that are akin to the porphyrins. The visible and near-UV absorption spectra of both systems are dominated by $\pi \rightarrow \pi^*$ electronic transitions and excitation in these bands produces strong resonance enhancement of Raman bands associated with ring vibrations. Several groups (Mayer *et al.,* 1973; Wozniak and Spiro, 1973; George and Mendelsohn, 1973; Galluzzi *et al.,* 1974; Tsai and Morris,

Table 5

Band (cm^{-1})	Oxidized derivatives[b]						
	Mb$^+$F$^-$	Mb$^+$H$_2$O	Mb$^+$HCOO$^-$	Mb$^+$SCN$^-$	Mb$^+$OCN$^-$	Mb$^+$OH$^-$	Mb$^+$N$_3^-$
I$_a$	—	—	160 (vwsh)	164 (vwsh)	162 (vwsh)	148 (vwsh)	—
I$_b$	—	194 (vwsh)	192 (vwsh)	—	187 (vwsh)	188 (vw)	187 (vw)
I$_c$	—	214 (vwsh)	214 (vwsh)	214 (vwsh)	214 (vwsh)	—	217 (vwsh)
II$_a$	252	249	252	249	255	255	254
II$_b$	270	272 (sh)	270 (sh)	271 (sh)	270 (sh)	271 (sh)	—
III	308	308	309	306	307	306	311
IV	347	346	347	341	341	347	342
	—	—	—	349 (sh)	354 (sh)	—	—
V	378	378	378	377	378	379	379
VI	416	411	412	414	413	413	411
VII	446	442	442	442	441	442	441
VIII	502	503	502	505	505	503	508
IX	556	552	552	555	555	556	554
X	679	678	678	679	680	678	677
lsb[c]	462	—	—	—	—	490	570

[a] Excitation wavelength is 441.6 nm. Adapted with permission from A. Desbois *et al., Bi... chemistry* **18,** 1510 (1979). Copyright 1979 American Chemical Society. The frequencies for MbC and MbNO differ from those given in the original. The new values (A. Desbois and M. Lutz, priva... communication) are corrected for the effects of photodissociation.

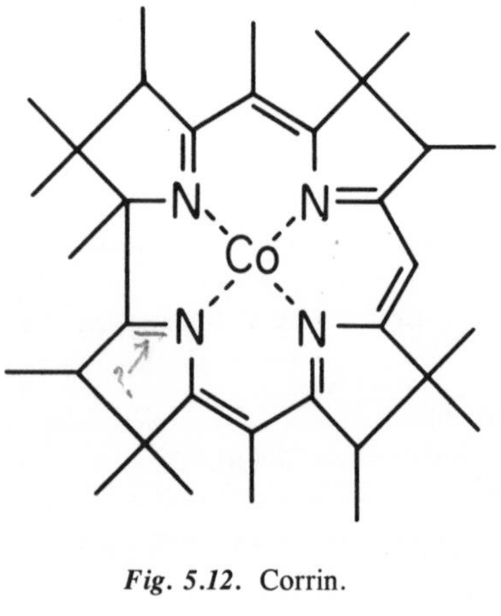

Fig. 5.12. Corrin.

ɔw-Frequency Vibrations of Various Myoglobin Derivatives[a]

Oxidized derivatives[b]		Reduced derivatives[b]			
Mb$^+$Im$^-$	Mb$^+$CN$^-$	Mb	MbO$_2$	MbNO	Assignment
–	—	109	—	—	—
3 (vwsh)	180	140 (sh)	—	—	—
16 (vwsh)	217 (vwsh)	—	212	—	—
55	254	222	256	256	ν[Fe–N(pyrrole)]
–	—	243 (sh)	—	—	—
09	306	305	307	305	—
47	347	345	345	344	Heme in-plane deformation
–	—	—	—	—	—
34	378	373	381	381	Heme in-plane deformation
13	413	408	410	410	ν[Fe–N$_\epsilon$(histidine)]
41	444	442	442	449	—
01	505	501	501	498	Heme deformation + [Fe–N(pyrrole)]?
53	554	551	549	551	—
79	677	675	678	677	Heme deformation
–	—	—	572	—	ν(Fe–sixth ligand)

[b] sh, shoulder; vw, very weak.
[c] Ligand-specific band.

1975; Salama and Spiro, 1977) have undertaken resonance Raman studies on the cobalamin family, including Co(II) and Co(III) derivatives and a variety of ligands in the axial positions about the metal. In a recent publication Salama and Spiro (1977) used a rapid-flow technique (similar to that described in Section VI,A,1, on the visual pigments) to obviate the problem of photolability encountered in earlier studies. The interesting conclusion to emerge from this work is that, unlike the porphyrins, the corrin ring-mode frequencies are insensitive to change in cobalt oxidation state or change in the nature of the axial ligand. Thus variations in these parameters appear to have little effect on the distribution of electrons within the corrin ring in the electronic ground state. However, a derivative containing Co(II) does show a large relative intensity change in bands at 1500 and 1600 cm^{-1} compared to some Co(III) derivatives, suggesting that the excited state properties are modified by a change in oxidation state. In the absence of frequency-variant corrin modes, identification of Co–ligand modes would provide an important "handle" on chemistry about the metal atom. Unfortunately, no bands attributable to Co–ligand stretching vibrations have, as yet, been reported.

IV. Chlorophylls

Chlorophylls are the principle agents in the photobiological conversion of light to chemical energy. The majority of chlorophyll molecules found in plants, photosynthetic bacteria, and algae act as bulk, or antenna, pigments that are instrumental in absorbing photons and transporting them to reaction centers. In the reaction centers, the chemical aspects of photosynthesis occur; there, other chlorophyll molecules and their pheophytin analogs bring about the conversion of light energy to electrical charge separation.

The importance of photosynthesis and the ability of RR spectroscopy to probe ground and excited electronic-state effects make the photosynthetic pigments prime candidates for RR spectroscopic studies. To date, however, only Lutz's group in Paris has undertaken extensive investigations of the RR spectra of the chlorophylls. Other researchers may have been deterred by problems of photolability and background luminescence. Lutz overcame the former problem by working at temperatures below 50°K. In this way he has provided detailed, and sometimes unique, information on some of the key pigments involved in photosynthesis. Although the antenna chlorophyll molecules exist in very complex biological environments, quite precise molecular information can still be obtained from the RR data. Lutz examined the antenna chlorophyll found in the photosynthetic membranes of various chloroplasts and algae (Lutz, 1972, 1977). Most of these systems contain both chlorophylls a and b (Fig. 5.13); however, by judicious choice of excitation wavelength in the Soret region, it was possible to selectively obtain RR spectra of either chlorophyll a or b. The tetrapyrrole skeleton found in the chlorophylls (Fig. 5.13) is closely akin to that present in heme proteins (Fig. 5.1), and, as in the case of heme RR spectra, the spectra of the chlorophylls contain a rich assortment of tetrapyrrole modes. By using ^{15}N isotopic substitution, Lutz was able to gauge the extent of C $\cdots$ C and C $\cdots$ N vibrational motions in the observed modes. He could, moreover, identify some lower frequency modes associated with the motion of the Mg atom by replacing ^{24}Mg with ^{26}Mg. As in the case of certain heme systems (Babcock and Salmeen, 1979), it is possible to identify features attributable to peripheral carbonyl substituents, and Lutz identified carbonyl stretches of the 9-keto carbonyl of chlorophyll a and b and the 3-formyl carbonyl of chlorophyll b. The possible mechanistic importance of the carbonyl groups and the fact that they are good group frequencies motivated a detailed study in this spectral region. The carbonyl region in the RR spectrum of antenna chlorophyll a in the intact algae *Botrydiopsis alpina*

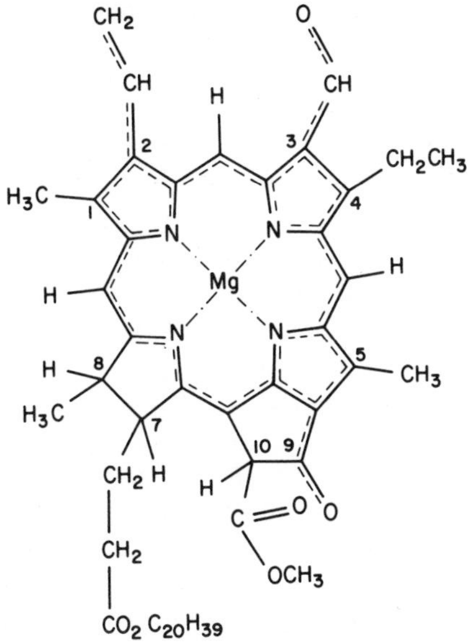

Fig. 5.13. Chlorophyll *b*. In chlorophyll *a* a methyl group replaces the formyl moiety at position 3.

differs markedly from that of the purified monomer in acetone or from those found for various aggregated chlorophylls in model systems. From this Lutz concluded that, *in vivo,* chlorophyll–chlorophyll interactions at the carbonyl groups do not occur, but, instead, the chlorophylls exist within chlorophyll–protein complexes. Additionally, using all the RR data, it was proposed that, regardless of biological species, antenna chlorophyll *a* exists in approximately five environments and chlorophyll *b* in two. Evidence was also found for fivefold coordination about the Mg atom in both *a* and *b* forms of antenna chlorophyll.

More recent studies have focused on the reaction centers rather than the bulk, antenna pigments. Since the number of chlorophyll molecules in the reaction centers are small compared to those composing the antennae, reaction centers must be isolated prior to spectroscopic examination. Functional reaction centers can be extracted from photosynthetic bacteria by the use of detergents. The reaction center, from several species of bacteria, consists of a three subunit protein of MW 70,000, containing four bacteriochlorophylls (BChl), two bacteriopheophytins (BPheo, a chlorophyll-like tetrapyrrole lacking a Mg atom), one carotenoid, two quinones, and one non-heme iron. Photon absorption within the reaction center leads to charge separation between two of the bound BChls. The latter absorb at 870 nm in organisms containing BChl a, and hence are designated P_{870}. It is thought that the electron thus produced migrates to one of the BPheos, absorbing specifically at 545 nm, then successively to each of the two quinones. By working near 30°K and by employing excitation wavelengths ranging between 350 and 600 nm it was possible to obtain selectively the RR spectra of the BChls, the BPheos, and the carotenoid (Lutz, 1980; Lutz and Kleo, 1976, 1979). The last molecule is discussed in Section V. Excitation of reaction centers from *Rhodopseudomonas sphaeroides* in the 528–535-nm or 545–550-nm regions, at 25°K or less, selectively yields RR spectra of each of the two BPheo molecules. Detailed examination of the RR spectra shown in Fig. 5.14 leads to the conclusion that the two BPheos are in very different environments. In particular, the band attributable to the 9-C=O group occurs at 1700 cm^{-1} in BPheo(535) but at 1678 cm^{-1} in BPheo(545). Thus it is likely that no strong interactions occur about the 9-C=O of BPheo(535), whereas in BPheo(545) the 9-C=O is strongly bonded to an unknown group. A further interesting facet of the spectra shown in Fig. 5.14 is the appearance of bands, marked with triangles or asterisks, which are either weak or absent

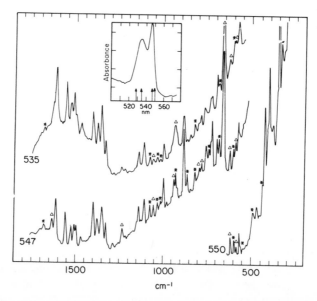

Fig. 5.14. RR spectra of bacteriopheophytins (BPheo) bound to photosynthetic reaction centers, 25°K, 535–550-nm excitation: Asterisks mark bands which are not observed for isolated BPheo; triangles mark bands strongly enhanced with respect to spectra of isolated BPheo. Inset: corresponding absorption spectrum in the 500–580-nm region. [From Lutz (1980).]

in the spectra of isolated BPheo. However, most of these features do appear in spectra excited near 360 nm in the Soret band. Lutz (1980) favors the explanation that the "new" bands result from BPheo–BPheo excitonic interactions in which the electronic excited states of adjacent chromophores are strongly coupled.

By excitation at slightly longer wavelengths, the RR spectra of the BChls come into prominence. When excitation wavelengths in the 580–610-nm region are used, the RR spectra of both the P_{870}^+ and the P_{870} can be detected. The oxidized P_{870}^+ form can be produced simply as a result of the laser excitation, whereas the P_{870} is obtained by using reducing agents such as dithionite. Additionally, the monomeric BChl a^+ cation radical can be produced in solution (Cotton and Van Duyne, 1978; Lutz and Kleo, 1979) and its RR spectrum compared to that of P_{870}^+ in the reaction center. The frequency shifts observed in the 1590- and 1375-cm^{-1} bands of P_{870}^+ compared to isolated BChl are at least as great as those seen in BChl a^+. This leads to the conclusion that the positive charge on P_{870}^+ is localized on a single BChl within the lifetime of the RR process because if delocalization onto a second BChl did occur on this timescale, then smaller frequency shifts would be expected. To observe carbonyl stretching modes from the BChl it is necessary to excite the RR spectrum with lines in the 350–360-nm region, within the Soret band. The resulting carbonyl profile is complex since, using excitation in this range, there are twelve putative C=O features from the BChls and BPheos in the reaction center. However, by comparing the observed spectrum with various model systems Lutz concluded that C=O $\cdots$ Mg or C=O $\cdots$ H$_2$O $\cdots$ Mg intrachlorophyll interactions are unlikely, a conclusion which casts doubt on proposed structures for the reaction centers which include these bonding features.

V. Carotenoids

The principal structural characteristic of carotenoids is a long, conjugated hydrocarbon chain. The extensive π-electron delocalization, seen in the structures of β-carotene and astaxanthin in Fig. 5.15, gives rise to intense visible absorption bands, and much of the pigmentation in nature, from flamingos to lobsters to carrots, is attributable to carotenoids. The biological functions of many carotenoids are unknown or conjectural, although in a few cases carotenoid function is understood (e.g., their role in camouflage or in protection of organisms from the deleterious effects of singlet oxygen produced during photosynthesis). For our purposes it is convenient to divide carotenoids into two classes: those which are highly photolabile and those which are not. The primary photolabile chromophore is retinal found in rhodopsin and bacteriorhodopsin, and these pigments are discussed in the following section. Here, the discussion is limited to those carotenoids which are relatively nonphotolabile.

Interestingly, the RR spectra of several carotenoids were reported, in 1932, by Euler and Hellstrom (1932). Coming just four years after C. V. Raman's initial report of the Raman effect, the work of Euler and Hellstrom represents the first RR study of a large truly biological molecule. During the 1950s the Raman technique was almost eclipsed by the upsurge in infrared spectroscopy; however, in Germany (see Behringer, 1967, and references therein) and Russia (see Shorygin, 1978, and references therein) some interest remained in the RR spectra of carotenoids because of the data they provided spectroscopists and theoreticians on Raman intensity enhancement. Then, in the late 1960s Rimai and co-workers in the United States began a series of investigations into the laser-excited RR spectra of carotenoids. The interest sparked by this work focused the attention of several laboratories on the potential of RR spectroscopy for elucidating the properties of biological pigments. Observations by Rimai and colleagues were stimulating because these scientists reported the RR spectra of carotenoids in intact plant tissue (Gill *et al.*, 1970) and the first studies of the visual pigments, in addition to the spectra of the purified chromophores from these materials (Rimai *et al.*, 1970c; Heyde *et al.*, 1971).

The essential RR spectral features of carotenoids are illustrated by astaxanthin which, like β-carotene, does not photoisomerize readily. Both molecules are shown in Fig. 5.15. Astaxanthin's intense $\pi \to \pi^*$ transition can be partially resolved into the vibronic components seen in Fig. 5.16 by going to low temperatures. The vibronic components 0–0, 0–1, etc., seen in Fig. 5.16, correspond to transitions between the $v'' = 0$ to $v' = 0, 1,$

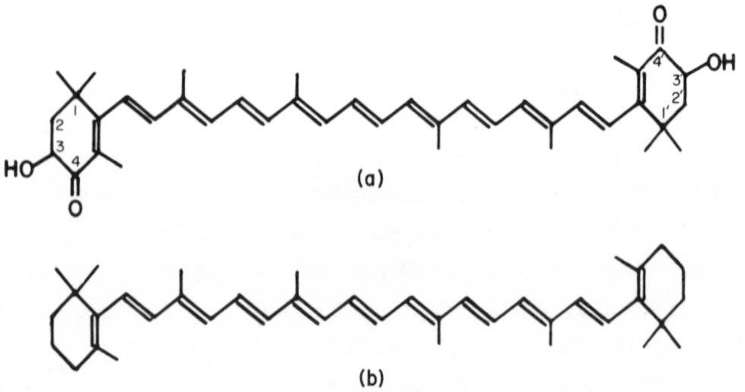

Fig. 5.15. (a) Astaxanthin, (b) β-carotene.

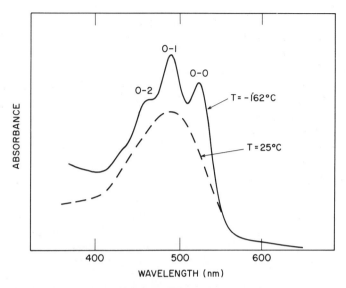

Fig. 5.16. Absorption spectra of astaxanthin at +25°C and −162°C. At low temperature the spectrum is partially resolved into its vibronic components.

etc., levels shown in Fig. 2.4, and each component contains contributions from every vibronically active normal mode. The RR spectrum of astaxanthin is shown in Fig. 5.17. As in the case of β-carotene (Rimai *et al.*, 1970b; Inagaki *et al.*, 1974; Sufra *et al.*, 1977), maximum intensity enhancement of the features at 1524, 1157, and 1006 cm⁻¹ occurs by excitation under the 0–0 transition. Additionally, for astaxanthin other lower intensity maxima are resolved at the 0–1 or 0–2 positions, when the Raman excitation profile is recorded at −162°C. However, at room temperature the loss of resolution in the absorption spectrum (Fig. 5.16) is accompanied by a loss in resolution in the excitation profile and the maxima at the 0–1 or 0–2 positions become ill-defined shoulders. This result demonstrates that the excitation profile does not always provide vibronic information hidden in an unresolved absorption profile (Salares *et al.*, 1976). In the RR spectrum shown in Fig. 5.17, the bands at 1524 cm⁻¹ and 1157 cm⁻¹ are associated with C=C stretching and C—C stretching modes, respectively, with the likelihood of coupling to C–C–H bending motions in both cases (Rimai *et al.*, 1973; Inagaki *et al.*, 1975). The C=C stretch, $\nu_{C=C}$, is an important marker band, and calculations by Rimai and co-workers led to the conclusion that its frequency is a genuine measure of stiffness of the C=C bonds and hence can be used to follow the degree of conjugation through the π-electron chain (Rimai *et al.*, 1973). For

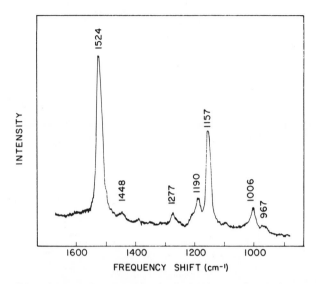

Fig. 5.17. RR spectrum of astaxanthin obtained using 4 mW of 465.8-nm excitation. Reprinted with permission from V. R. Salares *et al., J. Phys. Chem.* **80**, 1137 (1976). Copyright 1976 American Chemical Society.

carotenoids of different chain length and for a single carotenoid in a variety of solvents a correlation exists between $\nu_{C=C}$ and λ_{max}, the position of the absorption maximum for the carotenoid. As can be seen in Fig. 5.18 a plot of $\nu_{C=C}$ against $1/\lambda_{max}$ leads to a smooth curve (Heyde *et al.*, 1971; Rimai *et al.*, 1973). This shows, for example, that an increase in polyene chain length gives rise to increased delocalization of π electrons in the *ground state* (monitored by $\nu_{C=C}$) and a concomitant narrowing of the ground and excited state energy gap (monitored by λ_{max}).

A key advantage of RR, namely, the use of peak position to follow electronic ground state properties, has been utilized by Salares *et al.* (1977a, b) to identify strong excited-state effects. Certain carotenoids, e.g., astaxanthin (Fig. 5.15), form high-molecular-weight aggregates in aqueous solution. Aggregation leads to a large blue shift in λ_{max} from 480 to 400 nm, but little change in $\nu_{C=C}$. Thus the $\nu_{C=C}$ versus $1/\lambda_{max}$ correlation is not obeyed (Fig. 5.18). The unchanged $\nu_{C=C}$ shows that the ground state remains essentially unperturbed and that the shift in λ_{max} must be a property of the excited state. In fact, strong exciton coupling occurs in the aggregates; this has its origin in effects such as electronic transition dipole–dipole interactions between astaxanthin molecules and results in a massive perturbation to the excited states, but little change in ground state properties.

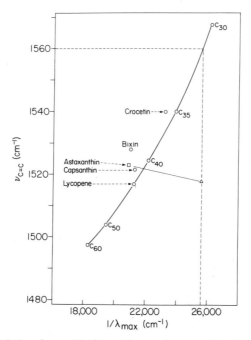

Fig. 5.18. Correlation of $\nu_{c=c}$ with $1/\lambda_{max}$. Monomeric astaxanthin (□) obeys correlation, aggregated astaxanthin (△) does not. [Adapted from Salares *et al.* (1977b).]

Astaxanthin has been used as an example in the previous discussion because of the author's interest in lobster pigmentation. The latter is entirely attributable to astaxanthin bound to various proteins. When a lobster is cooked, the astaxanthin-containing proteins denature, "unbound" astaxanthin is formed, and this "free" astaxanthin is responsible for the orange-red coloration of cooked lobsters. However, in the live crustacean protein–astaxanthin interactions bring about major absorption spectral shifts for the astaxanthin molecule. In one class of protein-astaxanthin complex a yellow protein is formed, while in a second kind of complex a blue protein results. It is a mixture of these that gives the live lobster its characteristic pigmentation. A yellow astaxanthin-bearing protein isolated from lobster shells was found, by a combination of RR and absorption techniques, to contain aggregated, excitonically coupled, astaxanthin molecules (Salares *et al.*, 1977a). Additionally, using RR *in vivo*, a film of yellow protein was found to envelop the shells of lobsters (Salares *et al.*, 1977a). Since the excitonic excited states of the aggregated astaxanthins could have the property of facile photon absorption and transmission, it was suggested that the entire lobster shell could act as a light-collecting dish to monitor low-light levels found at the bottom of the ocean (Salares et al., 1977a). Such a function could account for many of the temporal properties (e.g., seasonal migration) in the lobster life cycle. In this hypothesis, photons in the blue, green, and yellow parts of the spectrum are absorbed and passed along a chain of pigments with each successive pigment having a lower electronic energy gap than its predecessor. In other words, the photons are passed to pigments absorbing further in the red until, in the photoactive pigment, photochemistry takes place, leading to, for example, hormonal stimulation. Such a system would involve an as yet unidentified photoactive pigment absorbing further to the red of those presently identified in the lobster carapace. To begin a search for far-red absorbing pigments, the RR spectra of lobster shells were excited using infrared irradiation at 752.5 nm (Nelson and Carey, 1981). Figure 5.19 compares the spectra of lobster shell fragments taken with 647.1-, 676.4-, and 752.5-nm excitation. There are two notable features in these spectra. First, the value of $\nu_{C=C}$ at 1490.5 cm^{-1} in the 676.4-nm excited RR spectrum is the lowest value reported from an astaxanthin-like carotenoid. It is probably attributable to a pigment which has a λ_{max} to the red of 625 nm and which has not yet been isolated from the shell. The second unexpected feature is that the intensity of the 752.5-nm excited RR spectrum is higher (by a factor of 5–10) than expected on the basis of the presently known purified pigments. The unusual frequencies and intensities shown in Fig. 5.19 are inexplicable in terms of the known pigments, and further work may yield new perspectives on lobster pigment function.

The yellow protein isolated from lobster shell has its λ_{max} at 410 nm, and as mentioned previously, the shift from the value near 490 nm for free astaxanthin in organic solvents originates in astaxanthin–astaxanthin excitonic interactions. In contrast, the color of the blue proteins isolated from the shell is attributable to protein–astaxanthin interactions. The proteins appear to be blue because astaxanthin's λ_{max} is red-shifted to 610 nm. Prior to RR studies the favored model to account for the large red shift involved twisting about double bonds in the polyene chain (Zagalsky, 1976). However, the RR data eliminated this and several other mechanisms as possibilities (Salares et al., 1979). Instead, the findings of the RR investigation support a charge-polarization model. In this mechanism, charged protein groups and possibly hydrogen bonds in the astaxanthin binding site set up π-electron polarization and hence a dipole moment in the astaxanthin. Changes in the placement of charges about the astaxanthin change λ_{max} and thus account for the variations in λ_{max} observed for other astaxanthin-bearing proteins. Charge polarization induced by the binding site leads to a shift to the red in λ_{max} and a concomitant drop in

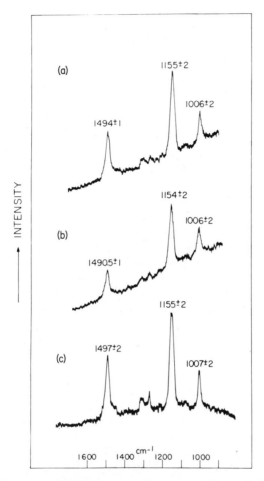

Fig. 5.19. RR spectra obtained from blue fragments of lobster shells using red and infrared excitation: (a) 647.1-nm, (b) 676.4-nm, and (c) 752.5-nm excitation. The fragments were immersed in buffer and rotated to prevent photodegradation. [From Nelson and Carey (1981).]

$\nu_{C=C}$. On this basis the $\nu_{C=C}$ observed at 1490.5 cm^{-1} in Fig. 5.19 must come from a very highly polarized astaxanthin in an, as yet, uncharacterized site. The charge-polarization model also explains how the λ_{max} of various visual pigments can be modulated (see Section VI). Furthermore, a charge-polarization model can explain the absorption and RR properties of various enzyme–substrate complexes discussed in Chapter 6, Section V.

In addition to their research concerning the chlorophylls and pheophytins in photosynthetic bacterial reaction centers, Lutz and co-workers
have also studied the carotenoid which is an intrinsic component of the
reaction-center assembly (Lutz *et al.*, 1978). The carotenoid spheroidene
is normally found in its all-trans conformation; however, in the reaction
center of *Rhodopseudomonas spheroides* the molecule binds in a cis conformation. This conclusion was reached by a detailed comparison of the
RR spectrum of the carotenoid *in situ* in the reaction centers with that of
15,15′-*cis*-β-carotene. The RR spectra of the all-trans and 15,15′-cis isomers of β-carotene are very different in the number, positions, and intensities of the major bands (Fig. 5.20). However, the agreement between the
cis isomer and reaction center is striking in all the regions of the RR
spectrum. The presence of the cis carotenoid was a general feature of the
reaction centers of all the species examined by Lutz *et al.* (1978). More
recently, these investigators have been able to add all-*trans*-spheroidene
to reaction centers from mutants containing no carotenoids and the RR
spectrum of the reconstituted complex showed that on binding the
spheroidene took on the cis conformation (Agalidis *et al.*, 1980).

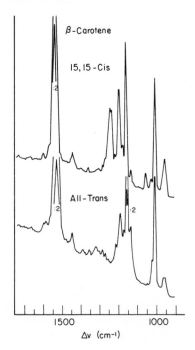

Fig. 5.20.   The RR spectra of 15,15′-*cis*- and all-*trans*-β-carotene: 30°K, 496.5-nm excitation. [From Lutz *et al.* (1978).]

The RR spectra of carotenoids are among the most intense and easiest to obtain. For these reasons trace amounts sometimes yield spurious peaks in the spectra from other "purified" materials. These properties also account for the appearance of carotenoid peaks in blood plasma (Larsson and Hellgren, 1974) and provide the opportunity for carotenoid assays in phytoplankton (Hoskins and Alexander, 1977). Finally, the occurrence of carotenoids in some membranes affords an opportunity to use their RR spectra as probes of membrane properties and this topic is discussed in Chapter 8, Section II,C.

VI. The Visual Pigments and Bacteriorhodopsin

RR spectroscopy is the most powerful method available for monitoring the photochemical transformations of the retinal chromophore (seen in Fig. 5.21). Absorption of a photon by retinal is the first step in two quite

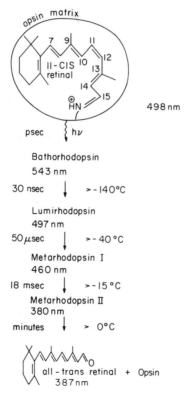

Fig. 5.21. The light-induced cycle of rhodopsin.

disparate biological functions. In one instance, in the eye, photon absorption leads to the visual process; in the second case, in certain bacteria, absorption is thought to energize a proton pump which sets up a proton gradient across the bacterium's membrane. The RR spectra can provide relatively detailed information on the various intermediates formed by these pigments, and in the past decade, several researchers have undertaken comprehensive studies on the retinal-based pigments using RR spectroscopy as their major tool. This important area has been the subject of recent reviews (Callender and Honig, 1977; Mathies, 1979), and here only a general overview is given with some mention of the ingenious methods which have been used to address the problems of identifying and characterizing the transient photo-induced complexes of retinal within its protein matrix.

A. Rhodopsin

The absorption of a photon by the photoreceptor protein rhodopsin leads to excitation of the retinal rod cell and forms the initial step in the visual response. Rhodopsin is a MW 38,000 membrane-bound protein which contains 11-*cis*-retinal covalently bound as a protonated Schiff base to a lysine side chain as shown at the top of Fig. 5.21. Following photon absorption, retinal is released from opsin in the all-trans conformation. Evidently a cis → trans isomerization occurs during visual excitation. However, the details of the photochemical pathway, the structure of the intermediates on the pathway, and the nature of the protein–retinal interactions within these intermediates are still unanswered questions. Using low-temperature absorption studies, Wald and co-workers (Yoshizawa and Wald, 1963) delineated the sequence of intermediates shown in Fig. 5.21. The RR work has concentrated on defining the molecular detail (with obvious emphasis on the chromophore) within each of the species making up the reaction sequence.

The first attempt at obtaining the RR spectrum of a visual pigment was reported by Rimai and co-workers (1970c). They cooled a bovine retina to $-70°C$, and, as shown in Fig. 5.22, were able to detect weak scattering in the 1550-cm^{-1} region. While the 1550-cm^{-1} feature was assigned to the C=C ethylenic stertch of a protonated Schiff base, Rimai and colleagues recognized that the conditions used to generate the RR spectra would give rise to a mixture of some of the intermediates shown in the reaction sequence of Fig. 5.21.

1. Controlling photolability

The inherent photolability of the visual pigments is, of course, fundamental to the process under study, and since Rimai's initial work several

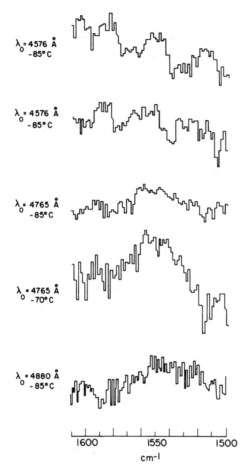

Fig. 5.22. An early attempt at obtaining a RR spectrum of retinal, *in situ,* within a bovine retina. [From Rimai *et al.* (1970c).]

ingenious methods have been used to define the species being probed by the Raman effect. An early approach utilized a double-laser-beam technique with one laser beam to "pump up" the population of a given transient and a second beam to monitor the RR spectrum. If the pump and probe beams are narrow "bursts" from pulsed laser sources, very short-lived intermediates can be monitored. In another approach, major progress in controlling the populations of intermediates was made by introducing very rapid flow methods. Two groups (Callender *et al.,* 1976; Mathies *et al.,* 1976) independently announced that they were able to obtain the RR spectrum of a single photolabile species by flowing rhodopsin at speeds of several hundred centimeters per second through a focused laser beam.

Both groups pioneering the rapid-flow technique derived a photoalteration parameter F, which is the fraction of rhodopsin isomerized while in the beam. The photoalteration parameters derived by the two sets of workers, are very similar; that of Mathies *et al.* (1976) being

$$F = 2.303 \times 10^3 \, P\epsilon\phi/N_0 lv$$

where P is the laser power (in photons per second), ϵ is the molar decadic extinction coefficient (in $cm^{-1}M^{-1}$), ϕ is the quantum yield for photo-isomerisation, v is the velocity of the stream (in centimeters per second), N_0 is Avagadro's number, and l can be approximated by the diameter of the laser beam. The rapid-flow technique has provided the excellent RR data on rhodopsin, isorhodopsin, and metarhodopsin I and II discussed in the following section.

2. Detailed consideration of rhodopsin RR spectra

The RR spectra of rhodopsin's photolytic intermediates are compared to the spectra of some model compounds in Fig. 5.23. With the exception of bathorhodopsin, the spectrum of each visual pigment bears close resemblance to the spectrum of the model shown alongside it. This comparison, taken with a detailed analysis, therefore identifies the conformation state of the retinal backbone and the state of protonation of the Schiff-base linkage in rhodopsin, isorhodopsin (a synthetic 9-*cis*-retinal pigment), metarhodopsin I, and metarhodopsin II.

The most intense feature near 1550 cm^{-1} in the spectra shown in Fig. 5.23 is the ethylenic double bond stretch, $\nu_{C=C}$. As in the case of the carotenoids shown in Fig. 5.18, $\nu_{C=C}$ correlates strongly with the position of the absorption maximum. For the rhodopsins, a red shift in λ_{max} is accompanied by a drop in the frequency of $\nu_{C=C}$, suggesting that a charge-polarization mechanism, similar to the one proposed for the astaxanthin bearing proteins (Section V; see also Salares *et al.*, 1979), also accounts for color changes among the visual pigments. The position of the C=N stretching mode provides the key to the state of protonation at this linkage. The C=N stretch occurs between 1655 and 1660 cm^{-1} in a protonated Schiff base, and near 1625 cm^{-1} in an unprotonated Schiff base. The features near 1657 cm^{-1} show that retinal is bound as a protonated Schiff base in rhodopsin, isorhodopsin, and metarhodopsin I. To date, a C=N mode near 1625 cm^{-1} has not been positively identified in metarhodopsin II to confirm the presence of an unprotonated Schiff base on this species. Instead, the identification relies on comparisons made in other regions of the RR spectrum (Doukas *et al.*, 1978). The conformation of the polyene backbone can be determined by analyzing the 1100–1300 cm^{-1}, or so-called fingerprint, region. The spectra of the model compounds (shown in

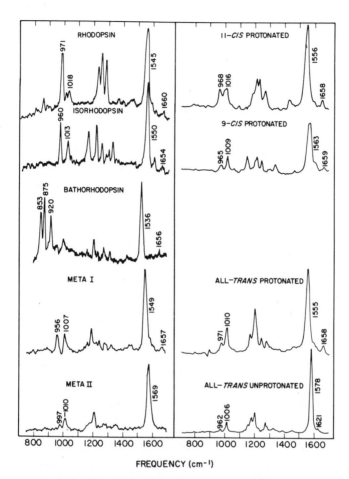

Fig. 5.23. RR spectra of rhodopsin and its photolytic intermediates (left-hand side) compared with the appropriate protonated and unprotonated Schiff-base derivatives of retinal (right-hand side). [From Mathies (1979).]

Fig. 5.23) demonstrate that the fingerprint region is very sensitive to various trans–cis isomerizations in the retinal chain. The similarities of the peak frequencies and intensities for rhodopsin and the 11-cis-model compound show that retinal is in an essentially unperturbed 11-cis conformation in the opsin matrix. Close similarities are also observed in the spectra of metarhodopsins I and II and all-trans protonated and unprotonated Schiff-base model compounds, respectively.

Establishing the conformation of bathorhodopsin has been particularly challenging. Oseroff and Callender (1974) were able to obtain the RR spectrum of bathorhodopsin by keeping rhodopsin at 77°K (thereby preventing photoisomerization from going beyond bathorhodopsin; see Fig. 5.21) and by using a "pump" laser beam to maximize the population of bathorhodopsin in a steady-state mixture of rhodopsin, bathorhodopsin, and isorhodopsin. They noted the poor correspondence between the bathorhodopsin and model-compound RR spectra (e.g., bathorhodopsin shows three unusually intense features at 853, 875, and 920 cm⁻¹) (Fig. 5.23). Subsequently, Eyring and Mathies (1979) (again using a dual-beam pump–probe method) were able to detect the Schiff-base vibration at 1657 cm⁻¹ and thus show that bathorhodopsin contains a protonated Schiff base. However, a major difficulty remained in determining the conformation of the polyene chain in bathorhodopsin and of correlating the conformation with the spectrum shown in Fig. 5.23. Several groups (Aton *et al.*, 1978; Sulkes *et al.*, 1978) favored a conformation involving various degrees of distortion of the polyene chain from that of the free all-trans isomer, and Warshel (1977) predicted, on the basis of calculations, that a strained all-trans conformer can have intense RR modes in the 800–950-cm⁻¹ region. The conformational issue has been tackled by Mathies and co-workers (1980) using some partially deuterated retinals. The visual pigment analogs formed with 11,12-dideutero and 10-monodeutero retinals demonstrated that the intense modes between 860 and 920 cm⁻¹ in the spectrum of bathorhodopsin (Fig. 5.23) are attributable to vinyl hydrogen out-of-plane bends. Using this assignment Mathies and co-workers (1980) could account for the intense features in calculations that assumed an all-trans conformation with small (∼20°) twists in the polyene chain.

B. Bacteriorhodopsin: The Purple Membrane Pigment

Certain bacteria growing on salt marshes contain a purple pigment in their membrane. The unique properties of this pigment were first described by Oesterhelt and Stoeckenius (1973), who showed that the purple pigment has some striking similarities to rhodopsin. The bacterial pigment contains retinal covalently linked by a Schiff base to a protein of MW 26,000, and the chromophore undergoes a series of photochemical transitions. For these reasons the purple membrane pigment is also known as bacteriorhodopsin. Bacteriorhodopsin has acquired considerable importance in the field of bioenergetics since it converts light to chemical energy, serving as a proton pump to create a proton gradient across the bacterium's membrane.

The photolytic cycle of bacteriorhodopsin, as it is presently understood, is shown in Fig. 5.24. Note that the purple membrane process is truly cyclic with a period of about 10⁻³ sec, whereas the recycling time for

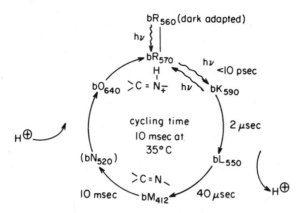

Fig. 5.24. The photochemical cycle of bacteriorhodopsin. Subscripts are the absorption maxima of the particular species in nanometers. [Adapted from Terner *et al.* (1977).]

the visual pigment is much longer and requires a final enzyme-mediated step to rejoin retinal to an opsin molecule (Fig. 5.21). Attention was given in the early RR work on bacteriorhodopsin to the characterization of the nature of the Schiff-base linkage. By comparing spectra taken in H_2O and D_2O the linkage in the bR_{570} complex was shown to be protonated (Lewis *et al.*, 1974; Mendelsohn *et al.*, 1974). Just as in the case of rhodopsin, a great deal of experimental skill has gone into obtaining the RR spectra of the intermediates in bacteriorhodopsin's cycle using variations and combinations of methods based on rapid-flow, pump–probe, double beams and pulsed-excitation sources (Lewis *et al.*, 1978; Terner *et al.*, 1979, and references therein). A generalization emerging from these studies is that the RR spectra of the various intermediates do not match the spectra of model compounds as closely as in the case of rhodopsin where the RR spectra of the intermediates can be modeled by the RR spectra of well-characterized isomers of retinal (Mathies, 1979; Aton *et al.*, 1977). Hence in the purple membrane there must be stronger retinal–protein interaction, giving rise to more strained retinal conformers, than occurs in rhodopsin. The protonated Schiff-base frequencies may be identified with some confidence by observing the frequency shift on making the N^+H to N^+D substitution, but, as is seen in Table 5.3, even these frequencies occur in a lower range compared to those in rhodopsin or model compounds. The interaction between the Schiff-base nitrogen and a protein group has been invoked to explain the low $C{=}N^+H$ frequencies seen in the purple pigment (Lewis *et al.*, 1978). The Schiff-base frequencies listed in Table 5.3 indicate that the bM_{412} intermediate is the only species containing an unprotonated Schiff base. Confirmation of an exchangeable

Table 5.3

Deuteration Effects on Bacteriorhodopsin Vibrations $(cm^{-1})^a$

Intermediate	Schiff base vibrations		800–1050- cm^{-1} region		Schiff base
	^1H_2O	^2H_2O	^1H_2O	^2H_2O	
bR_{560}^{DA}	1644	1622	1012	991	Protonated
	—	—	1012	979	—
	—	—	807	815	—
bR_{550}^{DA}	1644	1622	1012	991	Protonated
	—	—	807	815	—
bR_{570}	1644	1622	1012	979	Protonated
bK_{590}	1626	1616	984	986 or 1012	Protonated
bL_{550}	1647	1619	?	988	Protonated
bM_{412}	1623	1620	No effects	No effects	Not protonated
bO_{640}	1630	1616	?	992	Protonated
	—	—	?	965	—
	—	—	?	947	—

^{a} From Terner *et al.* (1979).

proton in each species in the cycle except bM_{412} is obtained by the observations (recorded in Table 5.3) that deuterium shifts occur in the 800–1050-cm^{-1} region (Terner *et al.*, 1979; Stockburger *et al.*, 1979).

In the fingerprint spectral region the bacteriorhodopsin species bR_{560} (dark adapted), bR_{570}, bK_{590}, bL_{550}, bM_{412}, and bO_{640} show very similar patterns between 1270 and 1390 cm^{-1} (Terner *et al.*, 1979). Since isomeric changes are almost certainly occurring among the intermediates, it seems that the 1270–1390-cm^{-1} range will not be of value in monitoring conformational changes in the purple pigments. This information could, in the future, be forthcoming from fingerprint bands between 1000 and 1250 cm^{-1}; however, a more sophisticated spectral analysis may be needed to unravel the spectra–conformation correlation. An additional feature of the bacteriorhodopsin problem is the existence of intermediates or putative intermediates not shown in the cycle of Fig. 5.24 (Marcus and Lewis, 1978; Stockburger *et al.*, 1979). Thus, a complete analysis of the photolytic transitions still presents a major challenge.

VII. Flavin Nucleotides

RR studies of flavin nucleotides are concerned with the isoalloxazine chromophore shown in Fig. 5.25. In riboflavin, the R group is D-ribitol; in

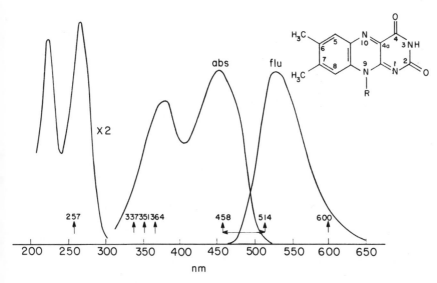

Fig. 5.25. Absorption and fluorescence spectra of flavin mononucleotide (FMN) in neutral aqueous solution. [From Nishimura and Tsuboi (1978).]

flavin mononucleotide (FMN), R consists of D-ribitol and a phosphate group. Flavin mononucleotide and flavin adenine dinucleotide (FAD, formed by linking FMN to adenine ribonucleotide by an anhydride bond), are important co-enzymes serving, for example, as prosthetic groups in certain classes of oxidation–reduction enzymes.

The development of the RR-flavin field has been hampered by the strong fluorescence associated with the isoalloxazine ring. However, this problem has been overcome in a number of ways. By taking into account the absorption and fluorescence profiles shown in Fig. 5.25, Nishimura and Tsuboi (1978) were able to avoid the fluorescence background by exciting the Raman or RR spectra of FMN and FAD with laser lines at 600.0, 363.8, 351.1, 337.1, or 257.3 nm. Compared to the 600-nm excited spectra, the RR spectra obtained by UV or near-UV excitation contained only a few features, suggesting that these RR spectra are limited by a low information content. Many more modes are RR active when the absorption peak near 450 nm provides resonance enhancement, but, under these conditions, steps must be taken to ameliorate the fluorescence background. Benecky and co-workers (1979) used solutions containing molar concentrations of KI which reduces the fluorescence background by a collision quenching process. Using this approach, RR spectra of FMN, FAD, and 7,8-dimethyl-1,10-ethylene isoalloxinium perchlorate were obtained and compared with the spectra of the FAD containing general fatty acyl-CoA dehydrogenase. In the latter complex, flavin fluorescence is quenched by interaction with the protein, and RR spectra can be recorded. Kitagawa and colleagues (1979a; Nishina *et al.*, 1978) have also used protein binding as a fluorescence-quenching tool and have reported detailed data (including several isotopically substituted molecules) on riboflavin binding to egg-white flavoprotein. A general conclusion arising from Kitagawa and co-workers (1979a) studies is that many of the features appearing in the RR spectrum are vibrationally mixed and contain contributions from two or more of the rings comprising the isoalloxazine system. An isoalloxazine mode near 1255 cm^{-1} is of particular interest since it can be used as a probe of 3N–H ⋯ protein interaction and Kitagawa *et al.* found that this mode involves large vibrational displacements of the C-2 and N-3 atoms with strong coupling to the 3-N–H bending mode.

Although technically sophisticated, CARS spectroscopy at present offers the only general solution to the fluorescence problem. A CARS spectrum of FAD in solution and bound to riboflavin binding protein is shown in Fig. 3.9. The CARS studies (Dutta *et al.*, 1977, 1978) again point to the feature near 1260 cm^{-1} as a band capable of yielding conformational information. This mode shifts to ~1295 cm^{-1} in the CARS spectrum in D$_2$O, to ~1250 cm^{-1} in riboflavin binding protein, and disappears altogether in glucose oxidase. Other bands undergo shifts of 3–5 cm^{-1} in both proteins.

An enzyme called old yellow enzyme (OYE) affords the interesting possibility of studying charge-transfer complexes involving FMN. Old yellow enzyme contains FMN and binds phenol derivatives with consequent inhibition of the enzyme and formation of a charge-transfer band (Fig. 5.26). Using excitation into the charge-transfer band, Kitagawa and co-workers (Kitagawa *et al.*, 1979b; Nishina *et al.*, 1980) were able to obtain strong resonance enhancement of both isoalloxazine and phenol modes. Excitation profiles show that intensity enhancement of the 1585- and 1550-cm^{-1} isoalloxazine features (known to involve vibrational displacements of the N-10 and C-4a atoms) is particularly remarkable. Depending on the phenol used to form the complex, various in-plane modes and a ring deformation mode (ν_{6a}) of the phenol were observed in the RR spectrum. Hence in the spectrum of old yellow enzyme complexed with pentafluorophenol (shown in Fig. 5.26), Kitagawa assigns the lines at 1517, 1475, 1310, and 454 cm^{-1} to the phenol modes ν_{19a}, ν_1, ν_{12}, and ν_{6a} respectively. The features at 1628, 1588, 1548, 1411, and 1359 cm^{-1} are attributed to isoalloxazine modes. The RR data suggest that the charge-transfer interaction takes place in a complex consisting of the phenol ring

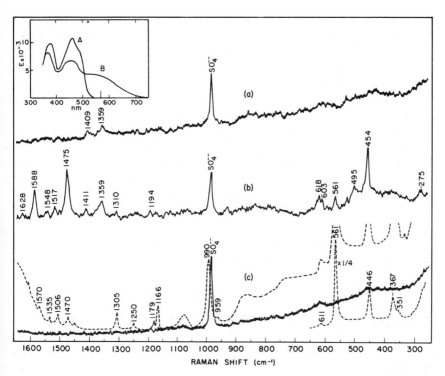

Fig. 5.26. RR spectra of (a) old yellow enzyme, (b) OYE–F$_5$Ph complex, and (c) F$_5$Ph: 568.2-nm excitation. Inset: absorption spectra of OYE(A) and OYE–F$_5$Ph(B). Reprinted with permission from T. Kitagawa *et al., J. Am. Chem. Soc.* **101,** 3376 (1979). Copyright 1979 American Chemical Society.

stacked over the isoalloxazine ring with the phenolate oxygen above the N-10 atom.

VIII. Metalloproteins

The statement made in the introduction to this chapter, that resonance Raman enhancement of bands arising from modes associated with the metal centers in proteins can arise only from excitation within the contours of charge-transfer transitions, needs some qualification. Recent studies of simple inorganic cobalt complexes suggest that in certain cases modest intensity enhancement of ligand–metal or ligand modes may be observed using metalloprotein transitions which are at least in part "d–d" (ligand–field) in origin, (Bosworth *et al.*, 1975; Chottard and Bolard, 1976; Salama and Spiro, 1978).

A. Hemerythrin, Tyrosinase, and Hemocyanin

The chromophore in each of these proteins is associated with a metal–metal dimer in the active site. None of the proteins contains a porphyrin-like heme group. Both tyrosinase and hemocyanin contain a binuclear copper active site and both proteins interact with molecular oxygen: hemocyanin acts as an oxygen carrier for molluscs and arthropods, whereas tyrosinase utilizes oxygen in the hydroxylation of monophenols and the dehydrogenation of o-diphenols. Hemerythrin is the oxygen-carrying protein of sipunculid worms and each subunit of hemerythrin contains two iron atoms in its active site. By implication it was thought that O_2 binds in the peroxide, i.e., O_2^{2-}, form at the Cu–Cu and Fe–Fe sites in these three proteins. RR spectroscopy was able to provide the first direct evidence of this.

In oxyhemocyanin an intensity enhanced oxygen–oxygen stretching mode was observed in the RR spectrum at 742 cm^{-1} (Loehr et al., 1974). The position is characteristic of peroxides, and assignment was confirmed by the shift to 704 cm^{-1} in $^{18}O_2$-substituted oxyhemocyanin. The 742-cm^{-1} mode is intensity enhanced by coupling to the ~570-nm absorption band assigned to a $O_2^{2-} \rightarrow Cu^{2+}$ charge transfer. A second intensity-enhanced mode, near 280 cm^{-1}, has been shown to be coupled to an absorption band near 340 nm and has been assigned to a Cu–imidazole stretch, (Freedman et al., 1976). Direct excitation into the 340-nm transition, produces intensity-enhanced spectra in the 100–400-cm^{-1} region with only a weak feature at 742 cm^{-1} remaining (Larrabee et al., 1977; Eickman et al., 1978). The UV-excited RR spectra of oxyhemocyanins from several sources are compared in Table 5.4 and show differences in detail. As in the earlier study (Freedman et al., 1976) assignment of the bands in the 200–350-cm^{-1} region to Cu–imidazole stretches is favored. The data in Table 5.4 suggest that there are at least subtle differences in the coordination geometries about the Cu^{2+} atoms within the oxyhemocyanins from various species.

Oxytyrosinase is formed by the reaction of tyrosinase with hydrogen peroxide in the presence of oxygen. Oxytyrosinase has absorption and RR spectra remarkably similar to those of the oxyhemocyanins (e.g., Table 5.4). As in the case of the hemocyanins, the energy of the 755-cm^{-1} band (Eickman et al., 1978) indicates that the oxygen is bound as the peroxide, and the shift to lower frequency by 41 cm^{-1} on $^{18}O_2$ incorporation is further confirmation since it is close to the theoretical shift of 43 cm^{-1} for a pure diatomic oscillator. The RR spectra in the 100–400-cm^{-1} region are also similar for oxytyrosinase and the oxyhemocyanins (Table 5.4; see also Eickman et al., 1978) indicating closely corresponding ligand character and geometry about the copper atoms.

Table 5.4

Resonance Raman Spectra of Oxyhemocyanins and an Oxytyrosinase[a]

Busycon canaliculatum	Limulus polyphemus	Cancer magister	Cancer irroratus	Cancer boreolis	Achatina fulica[b]	Neurospora crassa oxytyrosinase	Tentative assignment
119	—	—	—	—	—	—	δ_{NCuN}
170	180 (br)	180 (br)	110–240 (br)	199 (br)	—	184 (br)	δ_{NCuN}
226	223	218	—	217	—	218	ν_{CuN}
267	271 (sh)	262 (sh)	—	—	—	—	ν_{CuN}
286 (sh)	287	282	288	284	—	274	ν_{CuN}
315 (sh)	306 (sh)	308 (sh)	—	—	—	296 (sh)	ν_{CuN}
337 (sh)	338	333 (sh)	—	332	—	328	ν_{CuN}
749	752	744	—	748	752	755	ν_{OO}

[a] In cm^{-1}. Spectra obtained using 351.1- or 363.8-nm excitation except for *A. fulica*, where 514.5-nm excitation was used. The nitrogen atoms are from imidazole rings. Adapted with permission from N. C. Eickman *et al.*, *J. Am. Chem. Soc.* **100**, 6529 (1978). Copyright 1978 American Chemical Society.
[b] Chen *et al.* (1979).

For the binuclear-iron protein oxyhemerythrin, excitation into an absorption feature in the 550-nm region produces an intensity-enhanced oxygen–oxygen stretch at 844 cm^{-1}, shifting to 798 cm^{-1} on $^{18}O_2$ substitution (Dunn et al., 1973). The frequency is again characteristic of the peroxide O_2^{2-} form. Unlike hemocyanin, on visible excitation, a metal-oxygen mode is observed near 500 cm^{-1}. The RR spectrum of oxyhemerythrin in its natural state within erythrocytes is identical to that of the purified protein (Dunn et al., 1975). Hemerythrin also exists in an inactive oxidized state, methemerythrin, which reversibly binds any of a number of anionic ligands. The RR spectrum of the complex of sulfide and methemerythrin contains a single peak at 444 cm^{-1} and this has been assigned to an iron–sulfur stretch (Freier et al., 1979).

Mixed isotopes $^{16}O-^{18}O$ can be used to probe the symmetry of the oxygen binding sites, (Kurtz et al., 1976). The RR peak near 822 cm^{-1}, for the $^{16}O-^{18}O$ species oxyhemerythrin, comprises more than one feature. Thus the two oxygen atoms are in different environments, and using this it is possible to argue against models for the iron–oxygen site in which the oxygen atoms are bound to the iron atoms in an equivalent fashion. In contrast, for the copper protein oxyhemocyanin, mixed isotope studies by Thamann et al. (1977) show that the oxygen atoms are apparently equivalent.

B. Transferrins, Dioxygenases, and Other Fe(III)-Tyrosinate Proteins

Transferrins, dioxygenases, and other Fe(III)–tyrosinates all have Fe(III) as a prosthetic group and, as a result, exhibit visible spectra with absorbance maxima near 450 nm. The identity and stereochemistry of the protein ligands binding to Fe(III) have not been established with confidence. However, RR studies have shown unequivocally that one or more phenolate side chains bind to the Fe(III) site and that, as a concomitant, the visible absorption band may be associated with a phenolate → Fe(III) charge-transfer transition.

As their name implies, an important function of the transferrins found in serum is the transport of iron. Additionally, ovotransferrin, which apparently has an identical amino acid sequence to that of transferrin from blood serum of the same species, may be isolated from egg white. Both classes of protein avidly bind two Fe(III) atoms, and both have been studied by RR spectroscopy (Tomimatsu et al., 1973, 1976; Carey and Young, 1974; Gaber et al., 1974). Although recent reports using other techniques have begun to suggest some differences, it is thought that the binding sites about each Fe(III) are very similar, having tyrosine and histidine side chains and carbonate as favored ligands. The strong transferrin absorbance centered about 475 nm has been assigned to phenolate → Fe(III) charge transfer (Gaber *et al.*, 1974), and excitation into this band produces the RR modes seen in Fig. 5.27 near 1600, 1500, 1270, and 1170 cm^{-1}; all of which can be accounted for by phenolate ring modes (Tomimatsu *et al.*, 1976). The assignments are strengthened by

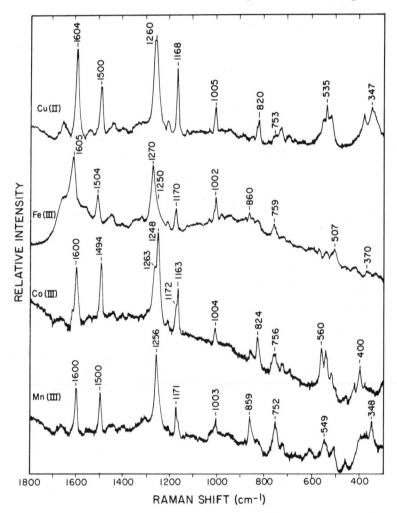

Fig. 5.27. RR spectra of aqueous solutions of ovotransferrin–metal chelates obtained using 488-nm excitation. Protein concentration was approximately $10^{-4} M$. Reprinted with permission from Y. Tomimatsu *et al.*, *Biochemistry* **15**, 4918 (1976). Copyright 1976 American Chemical Society.

comparison with Fe(III)- (Gaber et al., 1974) and Cu(II)-model compounds and by normal-coordinate analysis (Tomimatsu et al., 1976). Thus
the RR studies give clear direct evidence for phenolate (from
tyrosine) → Fe(III) ligation in transferrins. The absence of RR modes
from imidazole in the high-frequency region precludes immediate confirmation of histidine ligation. However, further analysis in the low-
frequency (50–500 cm^{-1}) RR spectrum, where metal–ligand modes are
expected, may prove informative. The Fe(III) atoms may be replaced by
Cu(II), Co(III), or Mn(III); several groups have reported RR spectra of
one or more of the substituted proteins (Siiman et al., 1974; Gaber et al.,
1974; Tomimatsu et al., 1976). The spectra in Fig. 5.27 are taken from a
particularly careful study by Tomimatsu et al. (1976), who compared the
RR spectra of the Fe(III), Cu(II, Co(III), and Mn(III) complexes of both
ovo- and human-serum transferrins. The spectra resemble that of the
Fe(III) protein in that they have intense phenolate features near 1600,
1500, 1270, and 1170 cm^{-1}. As with the case of iron–bicarbonate and
iron–oxalate ovotransferrins (Carey and Young, 1974), the subtle differences between the spectra in Fig. 5.27 illustrate the potential of the RR
technique for further comparative biochemical studies of the transferrin
systems.

Uteroferrin is a protein that bears a striking resemblance, in its RR
signature, to the transferrins (Gaber et al., 1979). Uteroferrin is an iron
containing glycoprotein purified from the uterine fluid of pigs. Excitation
within uteroferrin's visible-absorption band, centered at 545 nm, gives
rise to intense RR features at 1607, 1504, 1293, 1173, 873, and 803 cm^{-1}.
The similarities between the position of the first four bands and the corresponding peaks form the transferrins lead to the hypothesis that the
Fe(III) sites in the two proteins are similar (Gaber et al., 1979).

The dioxygenases are a further class of proteins that bear spectral
signatures resembling the transferrins and that have been the subject of
recent interest in the RR field. The intradiol dioxygenases, pyrocatechase,
and protocatechuate 3,4-dioxygenase, catalyze the cleavage of catechols
to cis,cis-muconic acids with the incorporation of molecular oxygen. The
enzymes have a high-spin ferric center in the active site and exhibit visible
spectra with absorbance maxima near 450 nm ($\epsilon = 3000$–$4000\ M^{-1}\ cm^{-1}$
Fe^{-1}. RR studies (Tatsuno et al., 1978; Keyes et al., 1978; Felton et al.,
1978; Que and Heistand, 1979; Bull et al., 1979) have assigned this absorbance to a phenolate (from tyrosine) → Fe(III) charge-transfer transition,
resulting in the enhancement of several phenolate vibrational modes, similar to those found for the transferrins and uteroferrin. The phenolate
modes from each class of iron–tyrosinate protein are compared in Table
5.5. Information on other ligands about the Fe(III) may, in the future,

Table 5.5

Resonance Raman Frequencies of Phenolate Ring Vibrations in Iron–Tyrosinate Proteins and Model Compounds[a]

	Phenolate frequencies (cm^{-1})			Reference
Ovotransferrin	1605	1504	1270	1172	Carey and Young (1974), Tomimatsu et al. (1976)
Serum transferrin	1613	1508	1288	1174	Gaber et al. (1974)
Lactoferrin	1604	1500	1272	1170	Loehr et al. (1981)
Uteroferrin	1607	1504	1293	1173	Gaber et al. (1979)
Protocatechuate 3,4-dioxygenase from *Psuedomonas aeruginosa*	1605	1505	1265	1176	Tatsuno et al. (1978), Keyes et al. (1978), Felton et al. (1978)
from *Psuedomonas putida*	1605	1504	1270	1175	Bull et al. (1979)
Pyrocatechase	1605	1505	1289	1175	Que and Heistand (1979), Que et al. (1980)
p-Cresol-Fe^{3+}, pH 7.0	1618	1488	1222	1180	Tatsuno et al. (1978)
p-Cresol, pH 14	1607	1490	1276	1176	Tatsuno et al. (1978)
Fe(EDDHA)$^{-}$	1600	1482	1286	1168	Gaber et al. (1974)
Fe(salhis)$_2$ClO$_4$, pH 7.0[b]	1625	1476	1337	1159	Que et al. (1980)
	1605	1452	1310	1132	Que et al. (1980)
Assignments	Ring stretch	Ring stretch	CO stretch	CH bend	

[a] Adapted with permission from L. Que et al., *Biochemistry* **19**, 2588 (1980). Copyright 1980 American Chemical Society. See Fig. 5.27 for other metal transferrins.

[b] Two sets of frequencies since it is an ortho-substituted phenolate.

come from analysis of the low-frequency (200–600 cm^{-1}) region of the RR spectrum (Bull *et al.*, 1979).

The enzyme–substrate and enzyme–substrate–O$_2$ complexes of the dioxygenase have also been probed by RR spectroscopy. Anaerobic substrate addition to the enzyme results in increased absorbance in the 600–800-nm region; RR studies in these complexes show that this additional absorbance arises from a catecholate-to-iron interaction, and provide evidence for substrate binding to Fe(III) (Felton *et al.*, 1978; Que and Heistand, 1979). The catecholate peaks observed in the RR spectra are similar to those seen in enterobactin and tris(catecholate)-ferrate(III) complexes (Salama *et al.*, 1978). The RR spectra of native pyrocatechase and its benzoate and phenolate complexes provide evidence for two distinct tyrosines coordinated to each Fe(III) (Que *et al.*, 1980). The evidence is in the form of two features in the 1270-cm^{-1} region which are interpreted as originating from the C–O vibrations of two differently coordinated phenolates. It is worth noting that the 1270-cm^{-1} band often appears broadened in Fe(III)–tyrosinate proteins, and, as in the transferrin spectra shown in Fig. 5.27, sometimes has a shoulder. Interestingly, for the dioxygenase it is found that the positions of the phenolate peaks shift in the ternary enzyme–substrate–O$_2$ complex of protocatechuate 3,4-dioxygenase, and, therefore, the tyrosine environment must be perturbed in this intermediate (Keyes *et al.*, 1979). Keyes *et al.* (1979) also formed the ternary complex using ^{16}O$_2$, ^{18}O$_2$, or ^{18}O–^{16}O, and since they found the RR spectrum insensitive to change in oxygen isotope, they concluded that no peaks could be assigned to dioxygen vibrations. Thus, unlike oxyhemerythrin, the optical spectrum of the complex has little or no contribution from peroxide → Fe(III) charge transfer.

C. Iron–Sulfur Proteins

These proteins are characterized by an iron–sulfur core in which the iron may be Fe(II) and/or Fe(III) and the sulfur is in the form of a thiolate ligand from cysteine and/or sulfide bridges. Iron–sulfur proteins are ubiquitous in nature, and the biological function of the small soluble proteins, where known, is redox coupling to enzymes engaged in a great variety of reactions. Their biological importance has led to many biochemical and chemical investigations.

Excitation into S → Fe charge-transfer transitions leads to intensity enhancement of iron–sulfur modes; rubredoxin provided the first example of a RR spectrum from a metalloprotein (Long and Loehr, 1970). Oxidized rubredoxin, containing one Fe(III) surrounded by four cysteine sulfurs in a distorted tetrahedral configuration, gave bands at 314 and 368 cm^{-1}

assigned to symmetric and nonsymmetric modes from FeS_4, and features at 150 and 126 cm^{-1}, possibly arising from bending modes (Long *et al.,* 1971).

Further studies have utilized selenium substitution for the acid-labile sulfurs in adrenodoxin to support band assignments (Tang *et al.,* 1973), and preliminary RR spectra have been reported for cubane-like Fe–S clusters (Tang *et al.,* 1975). With one possible exception (Blum *et al.,* 1977) the spectra published to date show intensity enhanced features only in the 100–400-cm^{-1} region, wherein Fe–S modes are expected, (Long and Loehr, 1970; Long *et al.,* 1971; Tang *et al.,* 1973, 1975; Siiman and Carey, 1980).

D. "Blue" Copper Proteins

"Blue copper proteins derive their name from the presence of one or more Cu sites having an unusually intense absorption band near 600 nm. Proteins with a blue copper site may or may not have oth.r types of copper centers in addition, and they function typically as electron carriers or oxidases in the reduction of O_2 to H_2O. The blue copper sites were without precedent in simple inorganic compounds, and their intriguing properties have spurred a host of chemical, biochemical, and spectroscopic studies.

Resonance Raman investigations were undertaken initially by two groups: Siiman *et al.* (1974, 1976) studied stellacyanin, laccase, plastocyanin, ascorbate oxidase, and ceruloplasmin; while Miskowski *et al.* (1975) presented data on azurin, plastocyanin, and ceruloplasmin. The data from the two groups were in good accord. By excitation into the 600-nm region, S $\rightarrow$ Cu charge-transfer band intense-RR modes are observed in the 200–500-cm^{-1} region (Fig. 5.28) wherein Cu–O, Cu–N, and Cu–S stretching vibrations are expected. The assignment of these bands proved to be challenging for two reasons. First, the coordination number and geometry about the Cu(II) atom were unknown, and, second, the amount of data for model compounds involving Cu–N and Cu–S ligation (available at the time of the first RR protein studies) was limited and proved to be inadequate. The first difficulty was resolved by X-ray analyses of the blue copper proteins azurin (Adman *et al.,* 1978) and plastocyanin (Colman *et al.,* 1978), which showed that the Cu sites in both proteins are probably four coordinate with ligation from a cysteine sulfur, a methionine sulfur, and two imidazole nitrogens from histidine residues. In azurin there is the possibility of glutamine as a fifth ligand (Adman *et al.,* 1978). Further study of the RR spectra of blue copper proteins (Tosi *et al.,* 1975; Ferris *et al.,* 1979) coupled with spectroscopic studies on a

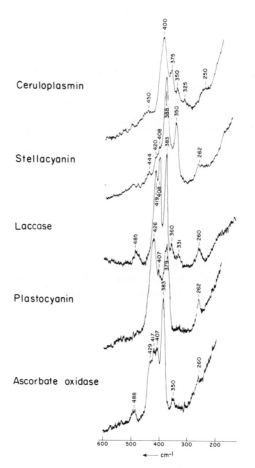

Fig. 5.28. RR spectra of some blue copper proteins, 647.1-nm excitation. Adapted with permission from O. Siiman *et al., J. Am. Chem. Soc.* **98,** 744 (1976). Copyright 1976 American Chemical Society.

variety of model compounds (Ferris *et al.,* 1978; Thompson *et al.,* 1979; Tosi and Garnier, 1979; Siiman and Carey, 1980) has led to the following interpretation of the RR spectra of blue copper sites. In general, the spectra show a feature near 260 cm^{-1} of weak or medium intensity and three or more intense bands between 350 and 430 cm^{-1} (Fig. 5.28) with only weak features occurring at higher cm^{-1}. The 260-cm^{-1} peak is thought to be ascribable to a Cu–S stretch where the S is from a methionine residue. However, in stellacyanin this band must have a different origin since stellacyanin does not contain methionine. The bands between 350 and 430 cm^{-1} are thought to be attributable to vibrationally mixed modes with contributions from Cu–S stretch (with the S of cysteine) and Cu–N stretch (with the N from the two ligated imidazoles).

In proteins containing both heme and Cu moieties the possibility arises that the absorption attributable to a blue copper center may be masked by the heme α band. Cytochrome *c* oxidase contains two heme groups and two Cu atoms, but excitation in the 600-nm region failed to detect any Cu–ligand vibrations from a putative blue-copper site (Bocian *et al.*, 1979).

Suggested Review Articles for Further Reading

General

Carey, P. R. (1978). Resonance Raman spectroscopy in biochemistry and biology. *Q. Rev. Biophys.* **11**, 309.

Carey, P. R., and Salares, V. R. (1980). Raman and resonance Raman studies of biological systems. *Adv. Infrared Raman Spectrosc.* **7**, 1.

Fawcett, V., and Long, D. A. (1976). Biological applications of Raman spectroscopy. *Mol. Spectrosc.* **4**, 125.

Spiro, T. G. (1976). Biochemical applications of resonance Raman spectroscopy. *Vib. Spectra Struct.* **5**, 101.

Spiro, T. G., and Gaber, B. P. (1977). Laser Raman scattering as a probe of protein structure. *Annu. Rev. Biochem.* **46**, 553. Contains comments on both the Raman and RR spectroscopy of proteins.

Spiro, T. G., and Loehr, T. M. (1975). Resonance Raman spectra of heme proteins and other biological systems. *Adv. Infrared Raman Spectrosc.* **1**, 98.

Hemes

Felton, R. H., and Yu, N.-T. (1978). Resonance Raman scattering from metalloporphyrins and hemoproteins. *In* "The Porphyrins" (D. Dolphin, ed.), Vol. 3, p. 347. Academic Press, New York.

Kitagawa, T., Ozaki, Y., and Kyogoku, Y. (1978). Resonance Raman studies on the ligand–iron interactions in hemoproteins and metallo-porphyrins. *Adv. Biophys.* **11**, 153.

Spiro, T. G. (1975). Resonance Raman spectroscopic studies of heme proteins. *Biochim. Biophys. Acta* **416**, 169.

Spiro, T. G. (1975). Biological applications of resonance Raman spectroscopy: Heme proteins. *Proc. R. Soc. London, Ser.* A **345**, 89.

Visual Pigments

Callender, R., and Honig, B. (1977). Resonance Raman studies of visual pigments. *Annu. Rev. Biophys. Bioeng.* **6**, 33.

Mathies, R. (1977). Biological applications of resonance Raman spectroscopy in the visible and ultraviolet: Visual pigments, purple membrane and nucleic acids. *In* "Chemical and Biochemical Applications of Lasers" (C. B. Moore, ed.), Vol. 4, p. 55. Academic Press, New York.

Metalloproteins

Kurtz, D. M., Shriver, D. F., and Klotz, I. M. (1977). Structural chemistry of hemerythrin. *Coord. Chem. Rev.* **24**, 145.

CHAPTER 6

Resonance Raman Labels

I. Introduction

Important biochemical interactions often occur at sites where a relatively small molecule is in contact with a highly specialized, local environment. Elucidation of these ligand–active-site interactions represents some of the key challenges of molecular biochemistry. When the ligand is colored, the chromophore provides an immediate "handle" that one may use to study the interaction by one or more of the techniques of optical spectroscopy, e.g., absorption, fluorescence, or circular dichroism. Intrinsic chromophores have, of course, been widely studied by RR spectroscopy and the detailed and sometimes unique information obtained from these systems is outlined in Chapter 5. However, many ligand–active-site interactions of interest to the biochemist do not contain an intrinsic chromophore which may be used as a RR probe of the interaction. To overcome this difficulty the RR-labeling technique was introduced (Carey *et al.*, 1972; Carey and Schneider, 1978). Resonance Raman labels are chromophores carefully designed to mimic natural biochemical components, and they are biologically active molecules. Just as in the case of naturally occurring chromophores these labels provide detailed vibrational and electronic spectral data from a biochemically important site.

154

The potential of the RR-labeling technique was demonstrated by a report that a good quality RR spectrum could be obtained of the dye methyl orange bound to bovine serum albumin (Carey *et al.*, 1972). An important conclusion from this research was that RR spectroscopy does provide detailed information on an extrinsic protein-bound chromophore under the biologically relevant condition of $\sim 10^{-5}\ M$ concentration in aqueous solutions. Since the initial report the RR-labeling technique has been extended to other systems which, for the most part, involve protein–ligand interactions. These include drug–enzyme and hapten–antibody complexes in which the drug or hapten molecules are the chromophoric RR labels. The information obtained from these studies is discussed in the sections that follow along with an outline of some enzyme-inhibitor complexes and a study of an enzyme carrying an irreversibly bound label in its active site. Each of these systems generally represents a stable 1:1 protein–ligand complex. However, by using chromophoric substrates it is possible to generate transient enzyme-substrate complexes and use the RR spectrum to probe the vibrational spectrum of the substrate during enzymic catalysis. This topic is discussed in Section V. Although, at present, the bulk of RR-labeling studies have concerned protein–ligand systems, the method has been extended to other important areas, such as nucleic acids and membranes. These applications are detailed in Chapters 7 and 8, respectively.

II. Drug–Protein Complexes and Other Enzyme-Inhibitor Systems

A. Sulfonamide Inhibition of Carbonic Anhydrase

Carbonic anhydrase is a zinc metalloprotein which catalyzes the reversible hydration of CO_2. The discovery that sulfonamides are strong inhibitors that bind to the active site with high affinity has proven to be of considerable therapeutic value. However, in spite of extensive physicochemical studies, including X-ray crystallographic analysis, the reasons for this high affinity are not entirely clear. The RR-labeling studies have provided several new insights and possible explanations. They show that the drug exists in the ionized form as $-SO_2NH^-$ when bound to the active site. In addition, they indicate that the geometry about the S atom in the bound, ionized form differs from that in the unbound ionized form.

Compounds **I**, **II**, and **III** (Fig. 6.1), chosen for their chromophoric properties and high affinity for carbonic anhydrase, were each studied (Kumar *et al.*, 1974, 1976) when bound to several of the available forms of

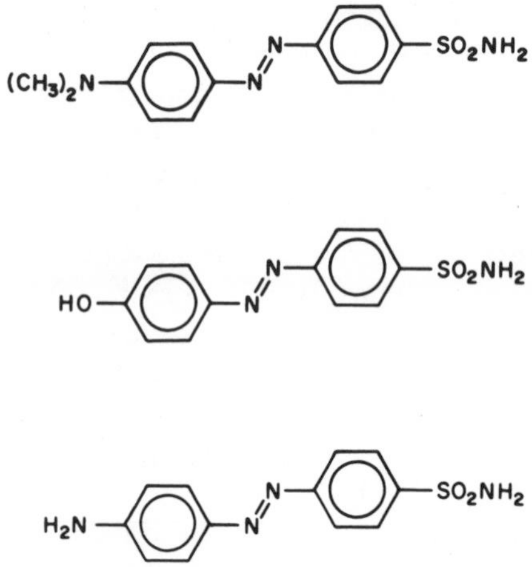

Fig. 6.1. Three azo-based sulfonamides used in RR studies of sulfonamides binding to carbonic anhydrase.

the enzyme. The ligands have intense absorption bands [probably mixed $\pi \to \pi^*$ and n $\to \pi^*$ transitions (Kumar and Carey, 1977)] between 380 and 480 nm (Fig. 6.2, inset). By excitation into these transitions, good quality RR spectra have been obtained in the 10^{-4}–10^{-5} M concentration range. The RR spectra and the tentative band assignments for compounds I, II, and III are given in Table 6.1. Figure 6.2 compares the RR spectrum of free aqueous II at pH 9.0 (Fig. 6.2c) with that of the sulfonamide in the active site at the same pH (Fig. 6.2a). Enzyme and solvent features are absent. Three alterations are observed in Fig. 6.2a: an increase in the relative intensity of the 1415–1388-cm^{-1} bands, a shift in the 1134-cm^{-1} band to 1138 cm^{-1}, and the appearance of a feature at 1123 cm^{-1} not seen in Figure 6.2c. The spectral perturbations attributable to binding are all slight, but analogous changes are seen in each of the spectra of compounds I, II, and III when bound to four different forms of carbonic anhydrase.

The spectral changes are attributed to an alteration in the structure of the sulfonamide on binding. The unbound form, at pH 9.0, is un-ionized. However, the bound form must possess an ionized sulfonamide residue $-SO_2NH^-$ because the spectra of this form in solution at pH 13 (Fig. 6.2b) and of sulfonamide bound at pH 9.0 are very similar. This finding is strengthened by the elimination of hydrophobic bonding or of changes in the azobenzene skeleton as sources of the spectral perturbations. Hydrophobic bonding effects were modeled by recording the sulfonamide spectra in solvents of varying dielectric constant and in the solid phase.

Table 6.1
Tentative Assignments (800–1600-cm⁻¹ Region) for Sulfonamide I, II, and III[a]

Assignment		Frequency[b]		
		Sulfonamide I	Sulfonamide II	Sulfonamide III
	8b	1593 w	1592 m	~1595 w
	19a	—	1488 w	—
Coupled	19b[c]	1446 m	1445 s	1464 s
	N = N	1419 vs	1417 vs	1428 vs
	19b[c]	1392 s	1384 vs	1398 s
	Ph–NMe₂	1370 m	—	—
	14	—	1328 w	—
	3	~1318 vw	1306 w	—
	Ph–O?	—	1250 w	—
Coupled	X sens 7a	1202 vw	1183 w	~1200 vw
	9a	~1160 sh	1157 w	~1175 sh
	Ph–N	1139 m	1134 m	1145 m
	18b	1097 w	1098 w	—
	5	921 m	924 m	928 m
	10a	826 w	843 w	846 w
	11	—	811 w	—

[a] Band positions are in cm⁻¹ and are for aqueous solution at pH ~9.0. Wilson's (1934) mode numbering is followed. From Kumar and Carey (1977).
[b] w, weak; m, medium; s, strong; vs, very strong; vw, very weak; sh, shoulder.
[c] Possibly both benzene rings contribute a 19b feature. It was suggested that the majority of ring modes come from the ring containing the –O⁻, –NH₂, or –N(CH₃)₂ group.

The spectra were found to be essentially insensitive to such changes. In particular, the intensity change and frequency shift observed on binding could not be reproduced by varying the medium. This does not mean that hydrophobic bonding is not an important source of binding energy, but rather that hydrophobic bonding does not produce the spectral changes. Additional insight is provided by vibrational analyses of azobenzene derivatives (Hacker, 1968; Kumar and Carey, 1977) which allow the prediction of the effect of change in conformation about the —N=N— bonds on the RR spectra. These studies indicate that there is no evidence for conformational changes in the azobenzene moiety on binding.

In addition to providing direct information about the nature of the bound form of the ligand, the spectra also suggest why the binding constants are so large. The key lies in the conformation of the –SO₂NH⁻ group which indicates that it may act as a transition-state analog. A feature unique to the bound spectrum and not seen in the spectra of the free –SO₂NH₂ or –SO₂NH⁻ forms is the new band near 1123 cm⁻¹ (Fig. 6.2a).

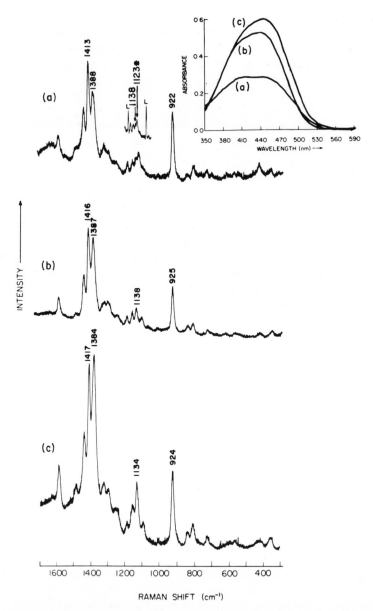

Fig. 6.2. RR spectra of $^-OC_6H_4N{=}NC_6H_4SO_2NH_2$ (a) bound to carbonic anhydrase, pH 9.0; (b) the free $-SO_2NH^-$ form at pH 13.0 in aqueous solution; (c) the free $-SO_2NH_2$ form at pH 9.0. Sulfonamide is $\sim5 \times 10^{-5}\,M$. Inset: the corresponding absorption spectra. (Reprinted with permission from K. Kumar *et al.*, *Biochemistry* **15**, 2195 (1976). Copyright 1976 American Chemical Society.)

It was proposed (Kumar *et al.*, 1976) that this new band results from a change in geometry about the S atom on binding over and above that resulting from ionization. This finding, together with a comparison of bond lengths and angles, led to the proposal that the $-SO_2NH^-$ group is mimicking the CO_2-OH^- transition state of the natural reaction for carbonic anhydrase. The similarity between the bound sulfonamide and a possible transition state for OH^- attacking CO_2 is shown in Fig. 6.3.

This example therefore shows that, using a very simple model, namely, the free aqueous $-SO_2NH^-$ form, there is strong evidence for sulfonamide ionization in the active site. Moreover, it was possible to rule out hydrophobic bonding and azobenzene structural changes as sources of the observed spectral changes. This contrasts sharply with other spectroscopic techniques (for example, fluorescence and absorption) where the observed data are more limited and cannot always definitively distinguish one cause to explain an observed effect.

Two further RR studies of a sulfonamide binding to carbonic anhydrase have been published (Petersen *et al.*, 1977, Carey and King, 1979). Both reports involve the sulfonamide Neoprontosil shown in Fig. 6.4.

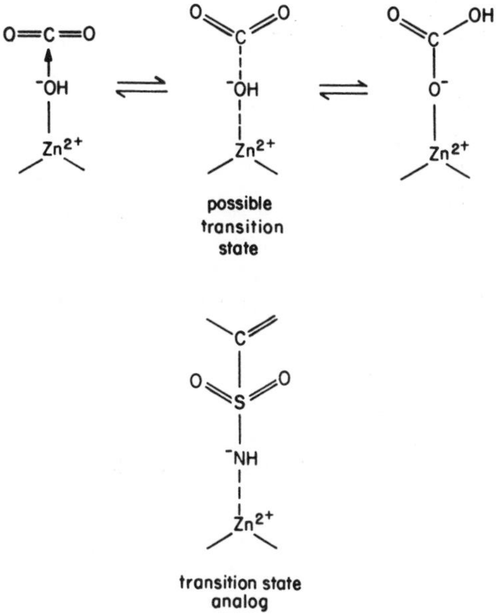

Fig. 6.3. Comparison of the binding of a sulfonamide with the proposed transition state for the natural reaction catalyzed by carbonic anhydrase. The zinc atom is in the enzyme's active site. (Adapted with permission from K. Kumar *et al.*, *Biochemistry* **15**, 2195 (1976). Copyright 1976 American Chemical Society.)

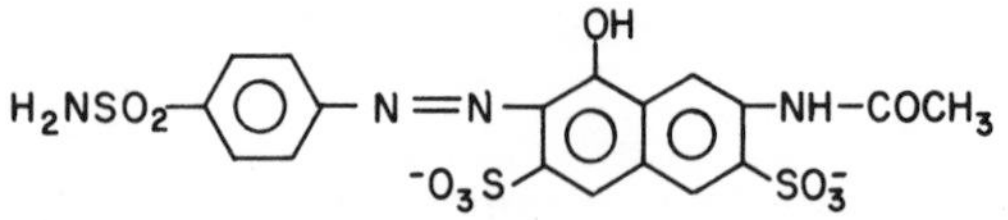

Fig. 6.4. Neoprontosil.

Compared to the sulfonamides **I, II,** and **III** shown in Fig. 6.1, Neoprontosil has less symmetry and more complex substituents, and these factors hinder spectral analysis. Additionally, there is a further complication centered around the pKs of the $-SO_2NH_2$ and $-OH$ groups on Neoprontosil. By using alkalimetric, spectrophotometric, NMR, and RR titrations, it was shown that the ionizations of these groups are "coupled" in the pH range 10.5–11.5 (Carey and King, 1979). The close proximity of the microscopic pKs for the $-SO_2NH_2$ and $-OH$ means that the distinction of Edsall *et al.* (1958) between microscopic and macroscopic pKs must be employed. The important consequence is that spectroscopic characterization of the separate $-SO_2NH_2$, $-O^-$, and $-SO_2NH^-$, $-OH$ species is impossible. Hence one cannot establish "standard" models for comparison with the RR spectrum of bound Neoprontosil; consequently, no conclusion can be drawn regarding the state of ionization of the $-SO_2NH_2$ group in bound Neoprontosil (Carey and King, 1979). In general, the case of Neoprontosil points to the need for caution in interpretation when the RR label contains two or more ionizable groups with similar pKs.

Larger differences in pKs did permit the study of the separate ionizations of the $-OH$ and $-SO_2NH_2$ groups in compound **II** shown in Fig. 6.1 (Kumar and Carey, 1977). Changes in the RR spectrum on $-OH \rightarrow O^-$ are more marked than those seen for $-SO_2NH_2 \rightarrow SO_2NH^-$. This reflects the fact (recognized by physical organic chemists) that the former ionization change is strongly transmitted to the π-electron system, whereas the $-SO_2NH_2$ ionization is not. In general, the Hammett substituent constants [the σ values (Hammett, 1970)] for a group located on a π-electron system give a good indication of the degree of participation of a substituent in the chromophore and its consequent effect on the RR spectrum. Thus substituents with large positive or negative σ values bring about marked changes in the RR spectrum compared to the unsubstituted compounds. Similarly, if a marked change in the Hammett σ constant occurs on ionization of a group [e.g., the σ (para) constant changes from -0.37 to -1.00 on $-OH \rightarrow O^-$], then a major change is expected in the appearance of the RR spectrum.

B. Methotrexate Binding to Dihydrofolate Reductase

The inhibition of the enzyme dihydrofolate reductase by methotrexate (MTX, Fig. 6.5a) is the source of the therapeutic effectiveness of methotrexate in the treatment of childhood leukemia and of several other cancers. Although MTX bears an overall structural similarity to the substrates of the enzyme (Fig. 6.5b), the affinity of the inhibitor for the enzyme is up to 10,000 times greater than that of the substrates. Part of this difference in affinity is thought to be that the MTX's pteridine ring becomes protonated on binding, whereas the substrate binds in the neutral form. Support for this concept comes from X-ray crystallographic work and various spectroscopic studies including Raman (Durig *et al.*, 1980) and RR (Saperstein *et al.*, 1978; Ozaki *et al.*, 1981) analysis. By noting the overall resemblance of the spectra of bound MTX and free protonated MTX, Saperstein *et al.* (1978) and Durig *et al.* (1980) argued that the pteridine ring of the bound drug is indeed protonated. The work of Ozaki *et al.* (1981) lent further support to this conclusion. Thus, in Fig. 6.6, the 350.6-nm excited RR spectrum of protonated MTX of pH 5.0 shows the closest resemblance to the spectrum of the bound drug. For protonated and bound MTX the correspondence between the 1269- and 1264-cm^{-1}, the 1373- and 1368-cm^{-1} and the 1414- and 1408-cm^{-1} bands is particularly noteworthy. The correspondence is not exact, however, and Ozaki *et al.* (1981) argued that enzyme–MTX interactions cause marked π-electron redistribution in the pteridine moiety over and above that attributable to protonation. The same researchers obtained data for the substrate folate bound to dihydrofolate reductase

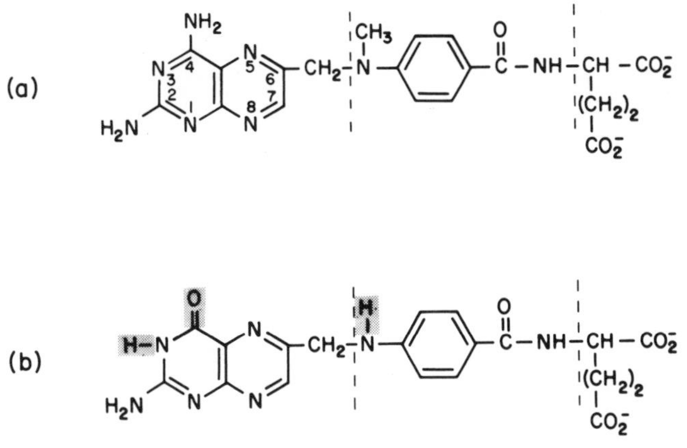

Fig. 6.5. The structural formulas for (a) the inhibitor methotrexate (MTX) and (b) the substrate folate.

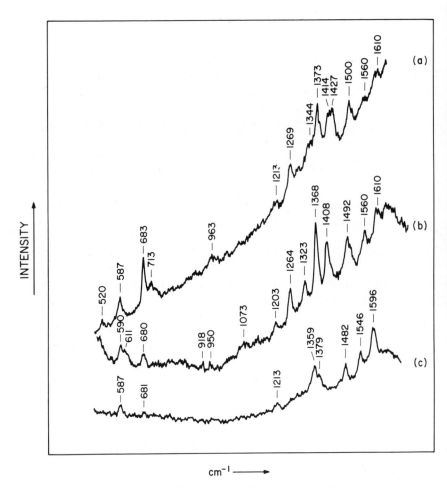

Fig. 6.6. The 350.6-nm excited RR spectra of MTX: (a) $5.5 \times 10^{-5} M$ in aqueous solution pH 5.0 (pteridine N1 protonated); (b) bound to dihydrofolate reductase from *Lactobacillus casei*; (c) aqueous solution pH 7.5 (pteridine neutral). (Reprinted with permission from Y. Ozaki *et al.*, *Biochemistry* **20**, 3219 (1981). Copyright 1981 American Chemical Society.)

and interpreted the data in terms of folate binding as a neutral species. A further observation of Ozaki and co-workers (1981) was that excitation with 350.6 nm resulted in a selective RR spectrum of the pteridine chromophore (the heterocyclic ring system in Fig. 6.5), whereas 324-nm excitation gave the RR spectrum of the *p*-aminobenzoyl chromophore. Thus, by switching excitation wavelengths, either the pteridine or the *p*-aminobenzoyl binding sites of MTX (or folate) were selectively probed.

C. Trypsin and Liver Alcohol Dehydrogenase– Inhibitor Complexes

Dupaix and co-workers (1975) studied the RR spectrum of 4-amidino-4'-dimethylamino azobenzene bound to trypsin. This chromophore is a competitive inhibitor of the enzyme with an inhibitor constant K_i of 2.3 μM at pH 6.08 and 15°C. The main features in the RR spectrum are similar to those listed for the azosulfonamide in Table 6.1 and the assignments of Dupaix *et al.* (1975) and Kumar and Carey (1977) (Table 6.1) are in good agreement. When 4-amidino-4'-dimethylamino azobenzene was bound to trypsin, no large frequency shifts occurred, but changes in the relative intensities of five bands at 1175, 1206, 1315, and 1608 cm^{-1} were observed. The lack of frequency shift of the bands assigned to $\nu_{N=N}$ and ν_{C-N} prompted Dupaix *et al.* (1975) to conclude that twisting of the phenyl group about the $-N=N-$ bonds did not occur in the bound inhibitor. A lack of twisting about this linkage was also observed for the azosulfonamides (Fig. 6.1) bound to carbonic anhydrase, but, as discussed in the next section, marked distortion does occur in some azo-based haptens.

In another early study of an enzyme–inhibitor complex, McFarland and co-workers (1975) obtained the RR spectrum of a metal-chelating agent bound to liver alcohol dehydrogenase. The latter enzyme contains two atoms of zinc per momomer of protein with one zinc atom being at or near the catalytic site. The chelating agent used was 2-carboxy-2'-hydroxy-5'-sulfoformazyl benzene (zincon) and the RR spectrum of this species changed markedly on binding to liver alcohol dehydrogenase. Since the spectrum of bound zincon was similar to that obtained from a well characterized zinc–zincon complex, McFarland and colleagues were able to deduce that binding to the enzyme's catalytic zinc atom had taken place.

III. Antibody–Hapten Complexes

Antibodies are key proteins in the body's immune defense system. The antibodies, which are molecules of MW $\sim$150,000, are synthesized in an organism in response to the presence of foreign substances known as antigens. To elicit antibody synthesis the antigen must be a high-molecular-weight substance, e.g., a protein or polysaccharide. However, antibodies to small and well-defined molecules can be made by linking the small molecule, called a hapten, to a protein carrier of high molecular weight. The hapten-bound protein is injected into an animal, and the antibodies specific for the hapten may be subsequently isolated from the animal's blood.

The first system of this kind studied by RR spectroscopy used the 2,4-dinitrophenyl (DNP) hapten and antibodies raised from rabbits (Carey *et al.*, 1973). Three haptens were used, two contained the DNP group linked to a substituted naphthalene ring via a —N≡N— moiety while in the third hapten the DNP group was bound in the ϵ position of a lysine molecule. In every case, many spectral changes, typified by the spectra shown in Fig. 6.7, were observed on formation of the antibody–hapten complex. On binding, features associated with —NO$_2$ symmetric stretching modes shifted indicating strong protein–nitro group interaction in all three complexes. For the two haptens containing the —N≡N— linkage, frequency or peak profile changes were observed for $\nu_{N=N}$ on binding. For example, $\nu_{N=N}$ for the hapten shown in Fig. 6.7 moves from 1417 cm^{-1} in the free form, to 1435 cm^{-1} in the bound form. These data were interpreted in terms of twisting occurring in the C—N≡N single bonds in the bound hapten which would reduce π-electron conjugation through the

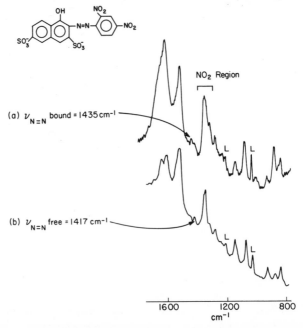

Fig. 6.7. RR spectra of a dinitrophenyl hapten (a) bound to rabbit antibody and (b) free in aqueous solution. Note the change in the $\nu_{N=N}$ and nitro symmetric stretching features on binding. An excitation wavelength of 457.9 nm was used. (Adapted with permission from P. R. Carey *et al.*, *Biochemistry* **12**, 2198 (1973). Copyright 1973 American Chemical Society.)

—N=N— and increase the double-bond character of the N=N linkage. Moreover, the observed broadening of the $\nu_{N=N}$ features was taken as evidence that binding in more than one conformation occurs and that the antibody combining sites vary slightly from antibody to antibody. This is an expression of the well-known fact that the antibody population is heterogeneous (i.e., the population contains a large number of slightly different antibody molecules all of which bind the hapten strongly).

The demonstration of a heterogeneous population of antibody binding sites, although of immunochemical interest, does present an additional hurdle to the process of spectroscopic interpretation. Therefore, in a subsequent study, Kumar *et al.* (1978) used two homogeneous antibody populations, termed MOPC 315 and MOPC 460 immunoglobulin As, derived from mouse tumors. The interpretation of the RR spectra of the DNP chromophores is not trivial, and a spectroscopic study was undertaken at the same time on a number of DNP based compounds (Kumar and Carey, 1975). In the biochemical study, the haptens used to form MOPC–hapten

Table 6.2

Dinitrophenol-Hapten Spectral Features and Effects of Binding to MOPC 315 and MOPC 460 IgA [a]

Hapten	Unbound	Bound to MOPC 315	Bound to MOPC 460
DNP–NH$_2$	1393 o,D	NC	NC
	1354 sh,o,D	NC	NC
	1338 p	1331	1325
	1280 o,D	1268	Broad
DNP–NHCH$_3$	1371 o	NC	NCν, Int ↓
	1340 p	NC	NCν, Int ↓
	1316 o,D	1308	1305
	1272 o,D	Broad	NC
ϵ-DNP–L-lysine	1375 o	1382	1384
	1338 p	NC	1327
	1315 o,D	NC	1302
	1275 o,D	Broad	Broad
DNP–N(CH$_3$)$_2$	1367 sh	NC	NC
	1330	1323	1322
	1310	1306	1303

[a] All measurements are in cm^{-1} and were taken in phosphate buffer. Notations: NC, no perceptible change; NC ν, Int ↓, no change in band frequency but a diminution in relative intensity; o, feature from the $-$NHR-o-NO$_2$ ortho structure; p, feature associated with the p-NO$_2$ group; D, ortho structure feature sensitive to deuteration; broad, feature too broad for exact wavelength to be determined; sh, shoulder. From Kumar *et al.*, 1978.

complexes were smaller and, therefore, more amenable to spectroscopic analysis than those used for the rabbit antibody studies. The four haptens used are listed in Table 6.2 and can all be regarded as 2,4-dinitrophenyl-aniline (DNP–NHR) derivatives. Isotopic substitution with ^{15}N and ^{2}H was used to assign features in the RR spectra of the free haptens. As a result of complex vibrational coupling in the molecules, the assignments are limited to associating features with either the para $-NO_2$ group or the ortho substituted $-NO_2$ and amino moieties. An ortho assignment means that the mode consists of a coupled vibration in which both the $-NH_2$ and its ortho $-NO_2$ partner participate. The assignments with the results of the antibody binding studies are shown in Table 6.2. The RR spectra of the bound haptens were obtained in the presence of KI to quench fluorescence; it was shown elsewhere that the KI has no effect on the spectra of the free chromophores (Kumar and Carey, 1975). The important conclusion from the data in Table 6.2 is that for both antibodies the nature of the R side chain in DNP–NHR appears to modify the interactions between the DNP chromophore and the protein. For example, the para feature in DNP–NH$_2$ shifts on binding to MOPC 315 or MOPC 460 indicating a strong protein interaction about the p-nitro group, but for DNP–NHCH$_3$ no frequency shifts are observed in the para feature; that is, at the level of resolution affecting RR peak positions there is not a unique binding site. These results emphasize that caution must be used in deducing binding-site structures from other spectroscopic studies (e.g., NMR and ESR) which sometimes assume that the DNP portion of several different haptens bind in the same fashion.

IV. Permanently Labeled
Sites—Arsanilazocarboxypeptidase A

Most RR labels are bound to the macromolecule in a noncovalent complex; however, it is possible to permanently link a RR-reporter group to a biologically active site. This type of label is an intruder in the sense that it does not have a natural counterpart. Nevertheless, a great deal of useful and precise information can still be gained regarding the active site. The prime example of a permanent RR label is the compound arsanilazocarboxypeptidase A whose chromophore is shown in Fig. 6.8. This derivative was originally synthesized by Vallee and co-workers, who succeeded in diazotizing a single tyrosine residue of carboxypeptidase A (Johansen and Vallee, 1971). This tyrosine, termed azoTyr 248 in the modified protein, was thought to be important for enzymic activity, although its placement with respect to the active-site zinc was controversial.

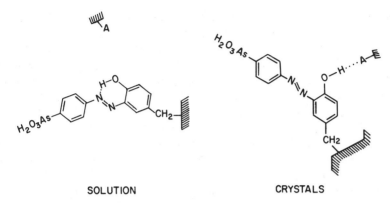

SOLUTION   CRYSTALS

Fig. 6.8. Schematic representation of the predominant conformations of the azoTyr 248 residue of azocarboxypeptidase in the solution and crystal phases. The hatched bar represents the protein and the letter "A" an undefined hydrogen-bond acceptor. (Reproduced with permission from R. K. Scheule *et al., Biochemistry* **19,** 759 (1980). Copyright 1980, American Chemical Society.)

Earlier absorption and circular dichroism (CD) studies had shown a marked dependence of the protein derivative properties on pH (Johansen and Vallee, 1975). Near pH 8.5, azoTyr 248 is coordinated to the active-site zinc atom, but at lower and higher pH values the complex dissociates to yield the protonated azophenol or azophenolate anion, respectively. RR titrations of the aqueous zinc azoenzyme confirmed the existence of three species of azoTyr 248 (Scheule *et al.,* 1977). The RR spectrum at pH 6.2 correlates well with those of model compounds (Scheule *et al.,* 1979) indicating that the azoTyr 248 is an aqueous environment. At pH 8.5, the spectrum correlates well with those of model azophenols forming complexes with zinc, this is in accord with the earlier absorption and CD data which led to the conclusion that the azoTyr 248 was coordinated to the active-site zinc. Scheule *et al.* (1980) favor a model in which this coordination is achieved by the phenolic oxygen and azo nitrogen of azoTyr with the azo group in the planar trans conformation. At pH 11, the coordination is broken, and azoTyr 248 exists as the ionized azophenolate species.

Detailed model studies have shown that the azoTyr chromophore can exist in two conformers. Both are essentially planar, but in one there is an intramolecular hydrogen bond (as shown in Fig. 6.8) while in the second conformer the $-OH$ group does not bind to any part of the chromophore. Each conformer has a characteristic $\nu_{N=N}$ and ν_{C-N} feature with intensities proportional to their concentrations. In arsanilazocarboxypeptidase A, the relative populations of these two isomers appears to change drastically on going from solution to the crystalline phase. Figure 6.9 compares the absorption and RR spectra of the azo enzyme in both phases at pH 6.2

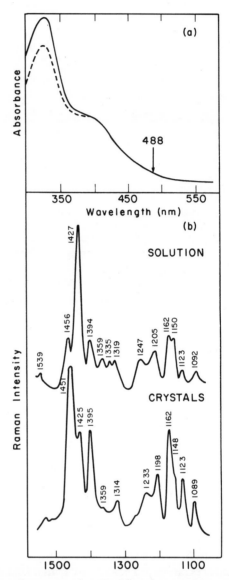

Fig. 6.9. (a) Absorption spectra of azoTyr 248 carboxypeptidase A in solution at pH 6.2 (solid line) and in a suspension of crystals pH 6.2 (dashed line). (b) RR spectra of the solution- and crystal-phase samples whose absorption spectra are shown in (a). (Reprinted with permission from R. K. Scheule *et al., Biochemistry* **19,** 759 (1980). Copyright 1980 American Chemical Society.)

where the azo chromophore is not chelated to the Zn atom. Although the absorption spectra are almost identical, marked changes in the relative intensities of some RR bands are seen. In the RR spectrum of the crystalline enzyme the bands near 1451 and 1123 cm^{-1} increase in intensity at the expense of the bands prominent in the solution spectrum near 1425 and 1148 cm^{-1}. Because the latter two bands are assigned to the intramolecularly hydrogen-bonded form (Fig. 6.8), this conformer must be virtually absent in the crystalline phase. Instead, the 1451- and 1123-cm^{-1} bands are assigned, on the basis of model studies, to a conformer in which there is a strong intermolecular H bond to an unknown protein acceptor designated A in Fig. 6.8.

V. Enzyme–Substrate Reactions

Each of the systems discussed in the preceding section of this chapter involves a stable $1:1$ complex between a chromophore and a protein molecule. It is also possible to monitor the vibrational spectrum of a substrate bound to the active site of an enzyme in an unstable enzyme–substrate complex (Carey and Schneider, 1974). The solution to the fundamental problem of how an enzyme recognizes its substrate and transforms it with inimitable specificity and efficiency is a central challenge in biochemistry. Since a vibrational spectrum is sensitive to the types of effects thought to be important during enzymolysis (e.g., bond strain, hydrogen bonding, and charge juxtaposition), the RR-labeling method has the potential for providing unique data and for discriminating between the various theories of enzyme catalysis. The research in this area has developed along two lines. Initially, substrates based on delocalized π-electron systems, such as cinnamic and furylacrylic acids, were used. The data on these systems can be unified and interpreted within the charge polarization model described in Section V,A. In a second approach, outlined at the end of this section, the dithioester ($-C(=S)-S-$) chromophore has been used to monitor the scissile bonds in natural enzyme–substrate reactions.

A. Substrates Based on Delocalized π-Electron Systems; the Charge-Polarization Model

1. Acyl papains and the charge-polarization model

A series of reactions involving the enzyme papain yielded results which initially were quite unexpected (Carey *et al.*, 1976, 1978). Papain is an enzyme of MW $\sim$23,000 which hydrolyzes amide or ester substrates by a transient acyl enzyme. In the latter, the acyl group of the substrate is covalently linked to a cysteine –SH side chain in papain's active site. A typical reaction is shown in Fig. 6.10. The acyl enzyme has a λ_{max} of 411 nm and can be purified from the reaction mixture by adjusting the pH and by subsequent column chromatography. The unexpected nature of the RR results is clear from Fig. 6.11. The RR spectrum of the enzyme substrate intermediate is distinct from that of the substrate or the product. The active site obviously produces drastic changes in the properties of the acyl residue. As can be seen in Fig. 6.11, the spectrum of the product is dominated by bands attributable to ethylenic and ring modes in the 1600-cm^{-1} region and a nitro feature near 1350 cm^{-1}. In contrast, the spectrum of the acyl enzyme shows a very intense peak at 1570 cm^{-1}. There is little evidence for peaks from the product in the spectrum of the intermediate or vice versa.

The dramatic change in the RR spectra of the cinnamoyl chromophore shown in Fig. 6.11 is thought to be attributable to the polarization of π-electrons in the bound cinnamoyl group (Carey *et al.*, 1978). A series of

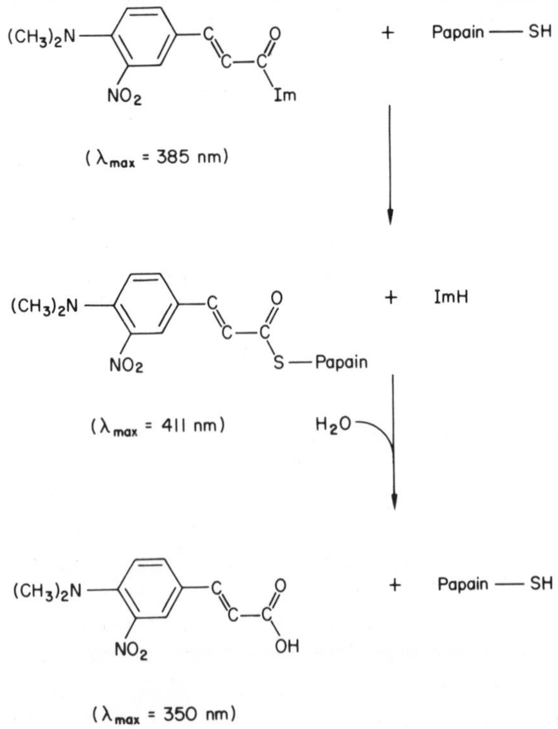

Fig. 6.10. Typical reaction involving the formation of a transient acyl papain.

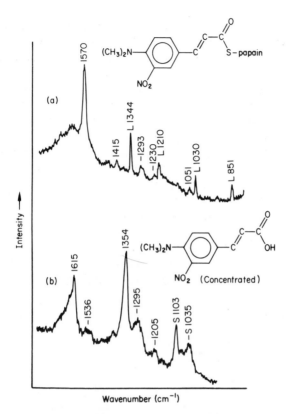

Fig. 6.11. Comparison of the RR spectra of the enzyme–substrate intermediate and the product. (a) RR spectrum of 4-dimethylamino-3-nitrocinnamoyl papain at 4°C, pH 3.0. (b) Pre-RR spectrum of 4-dimethylamino-3-nitrocinnamic acid. L and S denote laser plasma and solvent peaks, respectively. (Adapted with permission from P. R. Carey *et al., Biochemistry* **15**, 2387 (1978). Copyright 1978 American Chemical Society.)

model compounds, based on the imidazole esters of cinnamic acid, mimics the absorption and RR properties of the acyl papains studied. The imidazole ester of *p*-dimethylamino cinnamic acid is the principal model. It has a very strong electron-attracting group (imidazole) attached to the carbonyl and a very strong electron-donating group (*p*-dimethylamino) at the other extremity of the cinnamoyl skeleton. Acting in concert through the chemical bonds, these groups set up a highly polarized π-electron system. It has been proposed that essentially the same sort of electron polarization occurs in the acyl group bound at the active site. However, in the active site the polarization probably occurs indirectly by interaction of the acyl residue with protein groups.

The development of model compounds presents a good opportunity to test our ideas of electron polarization and, at the same time, to arrive at a semiquantitative estimate of the change in bond lengths occurring in the bound-acyl group of the substrate. The approach is to conduct precise structural determinations of the principal model compounds by X-ray crystallography, then to use the RR spectrum as a vector to carry this structural information to the enzyme-bound substrate. In simple valence–bond terms, the prediction that 4-dimethylaminocinnamoyl imidazole has a polarized π-electron system is illustrated by saying that structures of type **IV** make an important contribution to the true structure.

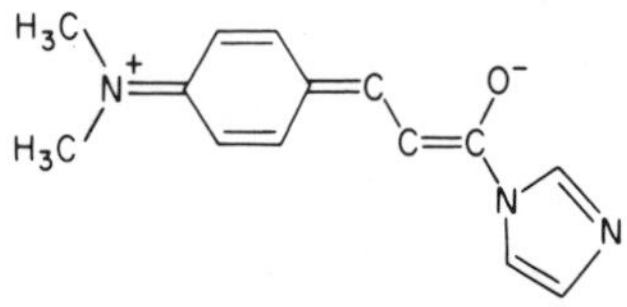

IV

This prediction is borne out by the X-ray structure shown in Fig. 6.12 (C. P. Huber, D. J. Phelps and P. R. Carey, unpublished work).

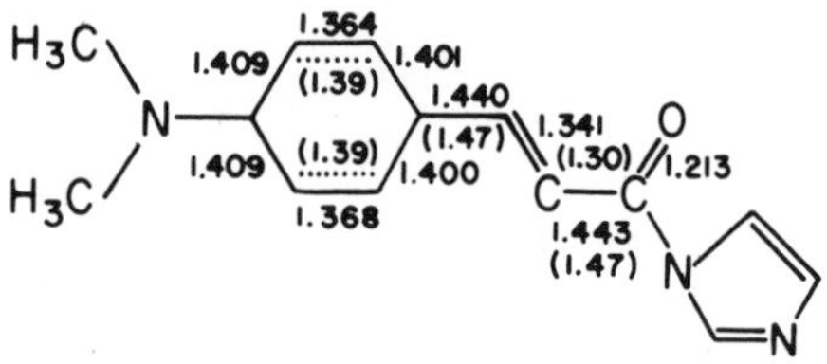

Fig. 6.12. Results of X-ray crystallographic analysis of compound **IV**.

Typical values for a cinnamic acid moiety, lacking powerful electron-donating or withdrawing groups, are given in parentheses. In comparison, the 4-dimethylamino derivative shows a tendency toward quinoid character as well as a significant shortening and lengthening of the ethylenic single and double bonds, respectively. In other words, the electron distribution implied by the valence structure of **IV** does approximate that in the model compound. The X-ray structure of the free chromophore is shown in Fig. 6.13 (C. P. Huber, D. J. Phelps, and P. R. Carey, unpublished work). Thus the precise structure of the free product and the structure for a crucial model compound that mimics the spectral properties of the chromophore in papain's active site are known. Therefore, by comparing Fig. 6.12 and Fig. 6.13 we arrive at a semiquantitative estimate of the structural changes occurring in the substrate on binding to the enzyme. Possible forces in the active site responsible for the large electron polarization in the acyl group are discussed at the end of the Section V, A, 2.

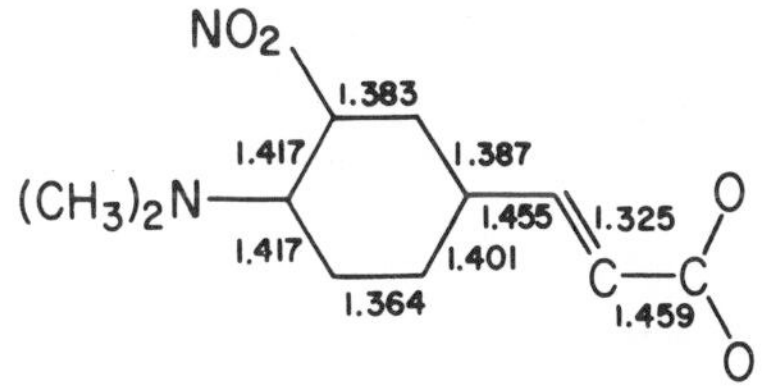

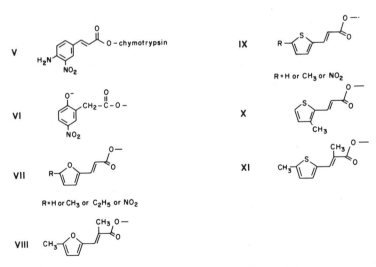

Fig. 6.13 Results of X-ray crystallographic analysis of the acid product, 4-dimethyl-amino,3-nitrocinnamic acid.

In general, it is not possible to obtain precise structural data by X-ray analysis for transient substrate–enzyme complexes. However, as is shown above, good quality RR spectra can be obtained from such transient species. Therefore, if model compounds can be found which mimic the spectral properties of the bound substrate the combined RR–X-ray approach offers considerable promise for delineating substrate conformational changes during enzymolysis.

2. Acyl Chymotrypsins

Compared to the acyl papains discussed previously the spectral changes observed for acyl chymotrypsins are small. Nevertheless, the changes can still be interpreted in terms of the charge polarization model. RR data have been obtained on each of the acyl chymotrypsins shown in Fig. 6.14, and the earliest spectra were obtained from the acyl groups V

Fig. 6.14. Acyl chymotrypsins that have been studied by RR spectroscopy.

and **VI** using excitation in the 450-nm region (Carey and Schneider, 1974, 1976). When reliable lasers operating in the near-UV became available, it was possible to examine the furylacryloyl derivatives shown in Fig. 6-14, which have absorption maxima near 350 nm. The absorption spectrum of 5-methylthienylacryloyl chymotrypsin [compound **IX** R = CH_3, Fig. 6.14] at pH 3.0 has a λ_{max} at 339 nm attributable to the acyl group and by excitation with 350.7-nm radiation the RR spectra shown in Fig. 6.15 were

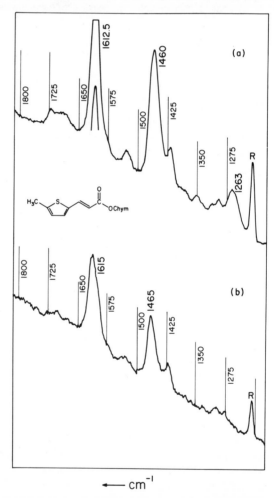

Fig. 6.15. The 350.6-nm excited RR spectrum of 5-methylthienylacryloyl chymotrypsin in (a) native form at pH 3.0 and (b) denatured by SDS. (Reprinted with permission from B. A. E. MacClement *et al., Biochemistry* **20**, 3438 (1981). Copyright 1981 American Chemical Society.)

obtained. Figure 6.15 compares the spectrum of the native acyl enzyme with that of the denatured form obtained by adding sodium dodecyl sulfate (SDS) detergent. The SDS "unravels" the secondary and tertiary protein structure, but the acyl group remains covalently bound. The features in Fig. 6.15 are all attributable to the acyl group and the most intense bands could be assigned to acryloyl or thienyl modes. Using spectra of this kind the following generalizations emerged for the group of compounds shown in Fig. 6.14, (MacClement *et al.*, 1981; Phelps *et al.*, 1981):

(i)   For the native acyl enzymes, the RR spectral profiles in the carbonyl region suggest that the acyl groups bound to Ser 195 in the active site adopt two conformations. These are characterized by having either strong hydrogen bonds to the $C=O$ (giving rise to the broad shoulder near 1700 cm^{-1} in Fig. 6.15a) or a nonbonding hydrophobic environment about the $C=O$ group (giving rise to the sharper band near 1725 cm^{-1} in Fig. 6.15a).

(ii)   In solution, the ester and acid analogs of the acyl group probably adopt more than one conformation about the acryloyl linkages. Therefore, the measured spectral parameters, such as the ethylenic double-bond stretching frequency $\nu_{C=C}$ in the RR spectra (corresponding to the 1612.5-cm^{-1} feature in Fig. 6.15), should be considered as a weighted mean, $\langle \nu_{C=C} \rangle$.

(iii)   For a series of compounds based on a given acyl group a correlation exists between $\langle \nu_{C=C} \rangle$ and the measured absorption maximum $\langle \lambda_{max} \rangle$. Changes in these spectral parameters are thought to be attributable to changes in electron polarization. One or more factors can change the polarization (e.g., H bonding at the $C=O$, charges or dipoles about the acyl chromophore, changes in bulk dielectric, and changes in conformation about the acryloyl's C–C bonds).

(iv)   Some acyl enzymes show a new band near 1260 cm^{-1}. This is seen clearly in the spectrum in Fig. 6.15a, but is absent in the spectra of denatured intermediates (Fig. 6.15b) and in the spectra of any model compounds, e.g., the acid or ester forms of the acyl group.

(v)   By changing pH it is possible to study the RR spectra as a function of enzyme activity. At pH 3.0 the acyl enzymes are perfectly stable because the enzyme groups required for catalytic activity are protonated and consequently ineffectual. However, at pH 7.0–8.0 the enzyme has optimum activity, and the acyl enzymes have half-lives of seconds. These unstable intermediates have been studied in flow systems, and a steady state of unstable complexes has been generated by mixing stable acyl enzyme at pH 3.0 with buffer at a pH near 7.0. On forming an active acyl enzyme there is a small increase in $\nu_{C=C}$ of ~3 cm^{-1} and a blue shift in λ_{max}

of 5–10 nm. These changes are ascribed to a reduction in π-electron polarization in the chromophore prior to deacylation which results from placing a negative or partial negative charge near the acryloyl carbonyl at active pH.

(vi)   Acyl enzymes formed from compounds **VIII, X,** and **XI** (Fig. 6.14) show unexpectedly low rates of deacylation, and the RR spectra of these intermediates indicate a perturbation in geometry of the acyl group on binding to the enzyme. It was proposed, therefore, that steric hindrance in the active site, caused by the substrates' CH_3 groups leads to low rates of deacylation.

(vii)   The spectrum of the native acyl enzyme shown in Fig. 6.15 was identical in H_2O and D_2O.

The question remains as to why the spectral changes observed for papain acyl enzymes are so much more dramatic than those observed for acyl chymotrypsin. An important difference between the two is that the acyl group is linked via an oxygen atom (from serine-195) in the chymotrypsin active site but via a sulfur atom (from cysteine-25) in the papain active site. Nevertheless, this change in covalent linkage can only account for part of the observed spectral differences. The principal cause is possibly connected with the α-helix dipole present in the papain active site. Cysteine 25, whose –SH side chain binds the acyl group during papain catalysis, is at the end of a portion of α-helical polypeptide chain. It has been proposed recently that α helices possess large dipole moments (Hol *et al.,* 1978), and this could account for the specially high degree of electron polarization (with the resultant spectral changes) observed for papain-bound acyl groups. In contrast, there are no α-helical fragments in the chymotrypsin active site, and the lack of major spectral changes for chymotrypsin acyl enzymes may be attributable to this fact.

3. Acyl glyceraldehyde-3-phosphate dehydrogenases

Chymotrypsin and papain are single-unit enzymes of MW $\approx 25,000$ with one active site per unit of enzyme. In contrast, glyceraldehyde-3-phosphate dehydrogenase is a tetramer composed of four equivalent subunits, each of MW 40,000, and each having its own active site. Such enzymes have important control functions because substrate binding in one active site is in some way transmitted to the remaining three sites with a consequent effect on their activity.

For the enzyme purified from rabbit muscle, it is possible by using furylacryloyl phosphate as a substrate, to bind nearly two furylacryloyl molecules per tetramer (Malhotra and Bernhard, 1968). The RR spectrum of this species appears to have two $\nu_{C=C}$ peaks; this, taken with absorption and kinetic data, led to the conclusion that the acyl enzyme exists as a mixed population of at least two forms (Storer *et al.*, 1981). However, the acyl enzyme prepared from glyceraldehyde-3-phosphate dehydrogenase from sturgeon appears to consist of a single population. The RR and absorption spectral properties were explained again in terms of the electron-polarization model. The furylacryloyl binds to a cysteine S in the enzyme active site and a series of model derivatives of type **XII**

XII

where X is H,N,O, or S were examined to aid spectral interpretation. When X is H, N., or O, $\nu_{C=C}$ shows a clear correlation with λ_{max}. However, when X is S, as in the acyl enzymes, the correlation breaks down. It has been proposed that this anomaly is attributable to empty sulfur $3d\pi$ orbitals, which, in certain conformations, may overlap with, and consequently change the electron distribution within, the ethylenic $p\pi$ orbitals.

B. Using Dithioesters to Observe the Point of Catalytic Attack

Resonance Raman studies of the acyl enzymes depicted in Fig. 6.14 provide new insights into enzyme catalysis and the molecular forces which operate in proteins. However, there are drawbacks associated with these systems since the substrates used are not the natural substrates for the enzymes concerned, and any mechanistic inferences based on the RR data must always keep this in mind. There is a further limitation in that intense RR features generally are not observed from the ester bonds composing the acyl enzyme linkage. Because these are the bonds undergoing catalytic transformation, it would be of great interest to monitor their vibrational modes during enzymolysis. These problems have been successfully overcome by the use of the dithioester chromophore which yields the RR spectrum attributable to just those bonds undergoing transformation in a "natural" enzyme–substrate complex (Storer *et al.*, 1979).

As mentioned previously, the catalytic hydrolysis of ester substrates by papain proceeds through the formation of an acyl enzyme in which a covalent linkage is formed to an active-site cysteine residue. Using absorption spectroscopy, Lowe and Williams (1965) were able to monitor this process using methyl thionohippurate as a substrate. They observed the transient appearance of a chromophoric intermediate with a λ_{max} at

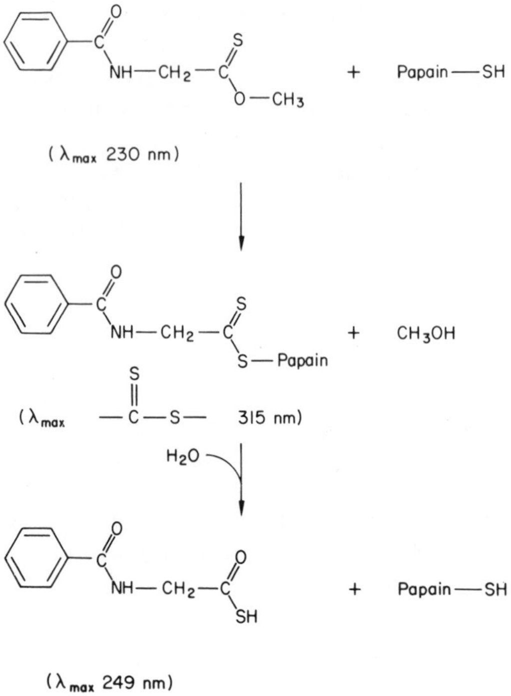

Fig. 6.16. Reaction involving the formation of a transient dithioacyl papain.

315 nm. Using kinetic evidence and absorption spectral comparisons with model compounds, Lowe and Williams inferred that this intermediate was a dithioacyl enzyme (Fig. 6.16). The intermediate has a λ_{max} at 315 nm, whereas the substrate and product both absorb below 250 nm. The RR data confirm that the intermediate is a dithioacyl enzyme. As can be seen in Fig. 6.17, by excitation with the 337.5-nm Kr^+ line, RR bands are observed at 1130 cm^{-1} and near 575 cm^{-1}. These features contain contributions from the stretching motions of the C=S and C—S—C bonds, and thus modes from the catalytically crucial bonds are observed in the RR spectrum. Moreover, the reaction scheme can be generalized to use, for example, a polypeptide sequence as substrate and any enzyme forming a transient linkage involving an active-site thiol. The "natural" reaction involves the formation of the

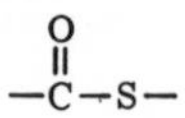

$$\begin{array}{c} O \\ \| \\ -C-S- \end{array}$$

group so the RR label now involves a single atom replacement-sulfur for oxygen.

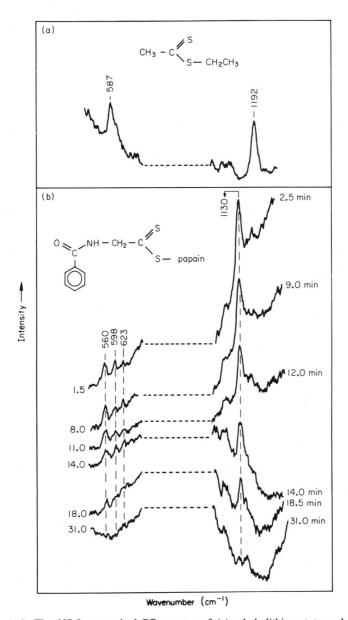

Fig. 6.17. The 337.5-nm excited RR spectra of (a) ethyl dithioacetate and (b) the enzyme–substrate intermediate, recorded in the C—S and C≡S stretching regions. For the dithio intermediate the numbers at the side of the spectra refer to the times in minutes after mixing of enzyme and substrate at which the 560- or 1130-cm^{-1} peaks were scanned through. (Reproduced from Storer *et al.*, 1979, by permission.)

As can be seen in Fig. 6.17 the RR spectrum of the dithioacyl enzyme disappears with time; in fact, the RR spectrum shows the same time dependence as the 315-nm feature in the absorption spectrum. Figure 6.17 also makes a comparison between the RR spectra of the enzyme-substrate complex and a simple dialkyldithioester. The two spectra are markedly different showing that the structure of the dithioester in the active site is substantially perturbed from that of the free chromophore in solution. The problem of understanding the conformation of the crucial bonds in the enzyme's active site now becomes one of interpreting the RR spectra typified in Fig. 6.17. Three modes of attack were used to overcome the problems of interpretation. First, the vibrational spectra of simple dithioesters, e.g., methyl dithioacetate

$$ H_3C-\overset{\overset{\displaystyle S}{\displaystyle \|}}{C}-S-CH_3 $$

and its analogs $D_3CCS_2CH_3$, $H_3^{13}CCS_2CH_3$, $H_3C^{13}CS_2CH_3$, and $H_3CCS_2CD_3$ were studied to provide insight into the normal mode structure of dithioesters (Teixeira-Dias *et al.*, 1982; Ozaki *et al.*, 1982). Second, a number of more complex dithioesters were investigated as "models" for the dithioacyl enzymes (Storer *et al.*, 1982). Third, series of related substrates and enzymes were set up in order to establish correlations between RR spectra, acyl group structures, and kinetic rate constants.

The dithioacyl enzymes examined and the corresponding dithioesters synthesized as models are shown in Fig. 6.18. Studies of the dithioester models shown in Fig. 6.18 reveal the presence of two conformers in solution. Since the RR spectrum of dithiohippuryl papain is shown in Fig. 6.17, we will consider its dithioester analog ethyl dithiohippurate. As can be seen in Fig. 6.19, the RR spectrum of this compound shows intense features in the "C—S stretching" (500–700 cm^{-1}) and "C=S stretching" (1070–1200 cm^{-1}) regions. In the latter range there are two features, at 1130 and 1177 cm^{-1}, whose relative intensities change drastically on going from CCl$_4$ to aqueous–CH$_3$CN solutions. All the complex dithioesters show this kind of behavior, and this, taken with variable temperature studies in the $-50°C$ to $+50°C$ range, indicates the presence of more than one rotational isomer in solution. One isomer gives rise to a characteristic peak near 1130 cm^{-1}, while the other isomer gives rise to a peak near 1180 cm^{-1}. A key finding is that the RR spectrum of one of the isomers (the one giving rise to the 1130-cm^{-1} band in Fig. 6.4) closely resembles the spectrum of the acyl enzyme. Therefore, delineation of the geometry of this isomer provides important information regarding the dithioester geometry in the active site.

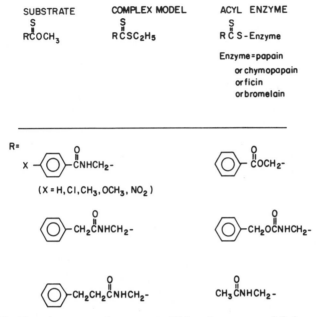

Fig. 6.18. The substrates used to generate dithioacyl enzymes and their corresponding dithioester model compounds.

Information on the exact structures of the rotational isomers was obtained by growing single crystals of the dithioester model compounds shown in Fig. 6.18. Combined X-ray and Raman studies on the crystals were used to establish structure–spectra correlations which, in turn, could be used to interpret the RR spectra of the esters in solution. Fortunately, the models shown in Fig. 6.18 crystallized into two main classes, each class bearing the Raman signature of one of the isomers found in solution. For example, the compound $CH_3C(=O)NHCH_2C(=S)SC_2H_5$ crystallized in an isomeric form which gave rise to the band near 1130 cm^{-1} (corresponding to the 1130-cm^{-1} band seen for the hippurate derivative in Fig. 6.19), while the crystalline form of p-$NO_2C_6H_4C(=O)NHCH_2$-$C(=S)SC_2H_5$ characterizes the isomer which yields the band near 1180 cm^{-1} (corresponding to the 1177-cm^{-1} band seen in Fig. 6.19). The structure of $CH_3C(=O)NHCH_2C(=S)SC_2H_5$ is shown in Fig. 6.20. The major difference between this structure and that of the p-$NO_2C_6H_4$ analog is, as far as the conformation near the dithioester group is concerned, a 150° rotation about the NCH_2–CS torsional angle. Hence it is inferred that the major difference in the two rotational isomers exhibiting the behavior illustrated in Fig. 6.19 is a 150° rotation about the NCH_2–CS bond. The

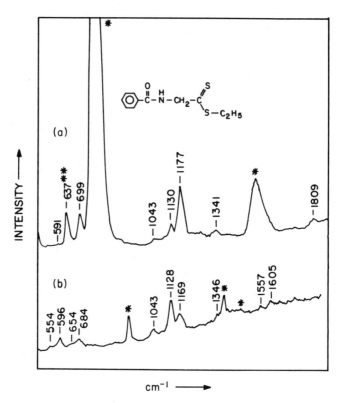

Fig. 6.19. RR spectra (324-nm excitation) of ethyl dithiohippurate (a) in CCl₄ and (b) in H₂O–CH₃CN mixture. Solvent peaks are denoted by an asterisk.

structure shown in Fig. 6.20 is particularly important because this form has a spectral signature similar to that of the corresponding acyl enzyme $CH_3C(=O)NHCH_2C(=S)S$—papain. Hence the conformation shown in Fig. 6.20 resembles the conformation of the acyl group in the native active site. When the acyl enzyme is denatured, the RR spectrum resembles that of the trace in Fig. 6.19b, showing that the acyl group exists in a mixed population consisting of both the conformers discussed previously. So it is apparent that the active site selects the conformation shown in Fig. 6.20 from a mixture of at least two thermodynamically accessible states.

Several of the dithioesters shown in Fig. 6.18 crystallized in forms whose spectra resemble those of the corresponding acyl enzymes. For example, the spectral characteristic near 1130 cm⁻¹ of the dithiohippurate seen in the solution spectra in Fig. 6.19 and the acyl enzyme spectra in Fig. 6.17 is present in the spectrum of crystalline ethyl dithiohippurate.

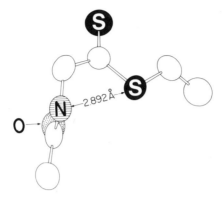

Fig. 6.20. The structure, derived by X-ray crystallography, of crystalline $CH_3C(=O)$-$NHCH_2C(=S)SC_2H_5$.

These additional structures provide further conformational information which can be used to establish structure–spectra correlations and delineate precise conformations of the scissile group during enzymolysis. Furthermore, the dithioester group provides a convenient optical "handle" for rapid-reaction kinetics such as the stopped-flow absorption technique, and, by this means, individual rate constants for the enzyme–substrate reaction may be obtained. The rate constants, combined with the structural data obtained by the RR–X-ray methods described previously have the potential to elucidate the relationship between kinetics and structure in enzyme reaction intermediates.

VI. Nucleic Acids and Membranes

The RR-labeling approach to nucleic acids and membranes is discussed in Chapter 7, Section V, B and Chapter 8, Section II, C, respectively.

Suggested Review Article for Further Reading

Carey, P. R., and Schneider, H. (1978). Resonance Raman labels: A submolecular probe for interactions in biochemical and biological systems. *Acc. Chem. Res.* **11**, 122.

CHAPTER 7

Nucleic Acids and Nucleic Acid–Protein Complexes

I. The Structure of Nucleic Acids

Two major classes of chainlike molecules underlie the functioning of living organisms. One class, the proteins, has been the subject of Chapters 4, 5, and 6. We now turn our attention to the second class, namely, the nucleic acids. Nucleic acids include deoxyribonucleic acid (DNA), which contains the hereditary message for each organism, and ribonucleic acids (RNAs), which help translate that message into each of the thousands of different protein molecules operating the living cell. Both DNA and RNA are long-chain polymers with alternating phosphate and carbohydrate groups and a purine or pyrimidine base attached to each carbohydrate moiety. There are four possible bases, and the identity of each RNA or DNA molecule depends on the sequence of the bases along the phosphate–carbohydrate strand. Figure 7.1 shows a segment of an RNA molecule; the bases adenine (A), uracil (U), guanine (G), and cytosine (C) are connected by glycosidic linkages to the ribose–phosphate backbone. In DNA, the 2′-hydroxyl group is replaced by hydrogen and the base thymine (T) occurs in place of uracil.

The structure shown in Fig. 7.1, taken with the sequence of bases along the chain, makes up the *primary* structure of nucleic acids. The *secondary* structure usually defines the relative orientations of the backbone and the base constituents. The secondary structure of a RNA

184

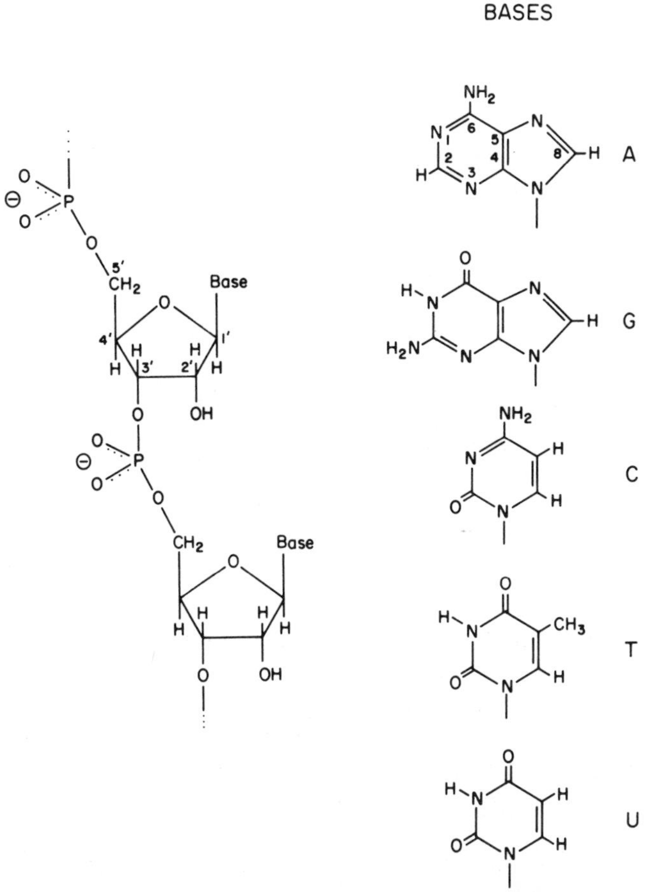

BASES

Fig. 7.1. Covalent structure of the nucleic acid ribopolymer chain (left-hand side) and the structures of the purine and pyrimidine bases (right-hand side).

molecule, known as transfer RNA (tRNA), is shown in Fig. 7.2. In some structures the planes of the bases lie one upon another, and the bases are said to be stacked. Complementary base pairing is an important structural feature. In DNA, T and A, usually on different strands, can form a dimer involving two hydrogen bonds between the bases, and G and C can form a dimer linked by three hydrogen bonds. Similarly, G–C and U–A pairs can form in RNA.

Tertiary structure refers to the complete three-dimensional structure of the nucleic acid, or to the relative orientations of regions of secondary structure. The double-helical structure of DNA is the classical example.

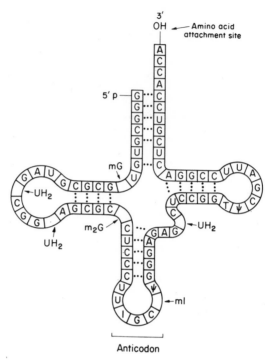

Fig. 7.2. The cloverleaf secondary structure of the RNA known as transfer RNA (phenylalanine) from yeast: tRNA$_{yeast}^{Phe}$.

In this case, two strands of nucleic acid are linked by complementary base pairs that may be represented by

$$-AACTGCTAGGT-$$
$$-TTGACGATCCA-$$

The double strand is twisted into a double helix with the base pairs forming the core and the deoxyribophosphate chain running along the outside of the helix. A model for a slightly distorted DNA is shown in Fig. 7.3. The distortion is attributable to complexation by certain histone proteins discussed in Section IV. Many important nucleic acid molecules do not possess the beautiful symmetry of double-helical DNA, and these less symmetrical molecules may have regions in which no obvious base stacking or base pairing occurs; nevertheless, they are held in a well-defined conformation by other specific intra- or intermolecular forces.

A molecule consisting of a single base linked to sugar is termed a nucleoside. By adding a phosphate, the unit becomes a nucleotide, and by

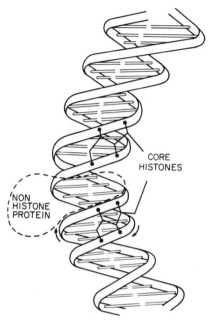

Fig. 7.3. A representation of the structure of DNA. The deoxyribophosphate backbones are represented by the two ribbons running the length of the double helix. The A–T and C–G base pairs are seen edge on and are represented by a pair of broken lines. The structure is slightly distorted by the binding of histone proteins as discussed in Section IV (Reprinted with permission from D. C. Goodwin *et al., Biochemistry* **18,** 2057 (1979). Copyright 1979 American Chemical Society.)

linking many nucleotides a polynucleotide is formed. Polynucleotides of well-defined sequence can be synthesized enzymatically or by organic chemists, and homopolynucleotides containing a single repeat unit [e.g., polyriboadenylic acid, poly(rA)] have been useful in establishing structure–Raman spectra correlations.

II. Polynucleotide Structure from the Normal Raman Spectrum

The normal Raman spectrum of a polynucleotide contains approximately 30 bands. Nearly all of these are attributable to purine or pyrimidine ring modes. However, two important features originating from the phosphate group have been identified; these are discussed in Section

II,B. The ribose or deoxyribose groups are poor Raman scatterers and do not contribute any intense or moderately intense features to the spectrum. The importance of the base ring modes has led to detailed analysis of their vibrational spectra. Moreover, because the bases have intense $\pi \rightarrow \pi^*$ electronic transitions in the 260–280-nm region, the interest has extended to the UV-excited RR spectra of the bases; this topic is discussed in Section V,A. The early research of the Raman spectroscopy of the purine and pyrimidine bases has been reviewed by Thomas (1975) and Peticolas (1975), while Parker (1971) has covered the IR investigations of these moieties. A benchmark paper in the Raman area was the extensive tabulation by Lord and Thomas (1967) of the Raman spectra of the bases, nucleosides, and the 5′-nucleotides. Their work is all the more remarkable because it was prelaser; the excitation source was the 435.8-nm line of a mercury lamp. In a more recent publication, Nishimura *et al.* (1978) reviewed their investigations of the normal mode analysis of the bases. This and similar studies (Delabar, 1978) rely heavily on isotopic substitutions to aid band assignments.

Before considering the use of ring modes to monitor the secondary structure of nucleic acids, we mention some other properties which can be probed by ring vibrations. The appearance of individual features attributable to, e.g., A,U,C, and G in a Raman spectrum affords the possibility of at least a semiquantitative estimate of the relative population of the bases, although possible hyper- and hypochromic effects (see Section II,A) must be taken into account. In adenine and guanine nucleotides the exchange of the 8-CH proton for a deuteron in D_2O occurs on the time scale of hours. Because this exchange leads to a marked change in the normal modes, Thomas and co-workers (Thomas and Livramento, 1975; Lane and Thomas, 1979) have been able to monitor the process in the Raman spectrum. The isotopically substituted analogs used in the work also yielded useful insights into assignments. Other researchers have reported the consequences of protonation of the ring bases; some ring vibrations show intensity and frequency changes that are diagnostic of the content of protonated or neutral species (O'Connor *et al.*, 1976a,b). Metal–nucleotide binding is another important area that has been addressed by Raman spectroscopy. Adenosine triphosphate (ATP) complexed with a divalent cation is a substrate for enzymes such as kinases and ATPases. Following the early work of Rimai and co-workers (1970a), Lanir and Yu (1979) have used the Raman spectrum to infer the mode of binding of ions such as Ca^{2+}, Mg^{2+}, Co^{2+}, Cu^{2+}, and Hg^{2+} to ATP. Evidence was found for metal–triphosphate coordination in every case between pH 3 and 12, but interactions with the adenine were usually metal and pH dependent.

A. Base Pairing and Base Stacking

In both DNA and RNA, certain purine and pyrimidine ring modes are sensitive to base-pairing and base-stacking interactions. Some of these features, and the general appearance of polynucleotide spectra, are illustrated by the spectrum (shown in Fig. 7.4) of poly(rA) · poly(rU), a double helix in which the adenine residues of poly(rA) hydrogen bond to the uracil residues of poly(rU). The system was investigated by Small and Peticolas (1971) and independently by Lafleur *et al.* (1972). As Fig. 7.4 shows, marked spectral changes occur on raising the temperature from 32° to 85°C. The spectral changes result from the thermal disruption of the helix with, at high temperature, the almost total loss of the base-pairing and base-stacking interactions.

Important spectral changes in Fig. 7.4 are intensity variations of certain ring modes and the radical changes in the carbonyl profiles between 1650 and 1700 cm^{-1}. In the latter region, in D$_2$O at 85°C, the carbonyl groups of U yield strong lines at 1661 and 1696 cm^{-1}, and at 32°C a single line occurs at 1688 cm^{-1}. Several lines, assigned to A and U ring vibrations, show appreciable intensity loss on helix formation at 32°C. This is

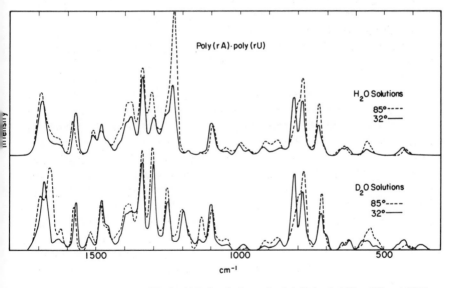

Fig. 7.4. Raman spectra of H$_2$O and D$_2$O solutions of poly(rA) · poly(rU) at 32° and 85°C. The background of Raman scattering by the solvent has been subtracted from each spectrum. [From Lafleur *et al.*, (1972).]

called *Raman hypochromism,* and intensity changes of this kind were first observed by Aylward and Koenig (1970) and independently by Tomlinson and Peticolas (1970) for polyadenylic acid. The phenomenon has a direct correspondence with the hypochromism observed in the UV-absorption spectrum, and it is believed to arise from electronic interactions between vertically stacked bases (Tomlinson and Peticolas, 1970; Painter and Koenig, 1976a). A few lines, however, show modest intensity gain in the 32°C spectrum (e.g., the mode near 1570 cm^{-1} increases in intensity with increasing structural order), and this effect is termed *Raman hyperchromism.*

Similar effects are seen in the spectrum of the poly(rC) · poly(rG) complex. Again there are several hypochromicities and a notable hyperchromic band at 670 cm^{-1} attributable to a G ring mode. In poly(rC) · poly(rG), however, the intensities of lines in the carbonyl region are weaker, by a factor of 6–8, compared to the carbonyl modes in the A · U helix.

At this stage, it is tempting to make the assumption that in a RNA containing a heterogeneous sequence of bases the carbonyl region may be used to quantitate base pairing, and the purine and pyrimidine ring modes may be used to form a semiquantitative picture of base stacking. However, difficulties arise from the fact that the profile in the carbonyl region and the degree of hypochromism are both dependent on base sequence. Moreover, some hyperchromic or hypochromic lines may reflect base-pairing as well as base-stacking interactions. For example, Lafleur and co-workers (1972) have suggested that bands above 1450 cm^{-1} in poly(rU) · poly(rA) may involve vibrations of H-bonding groups in the bases, and this could provide a mechanism by which base-pairing interactions perturb bands which, nominally, are attributable to base ring modes. The dependence on base sequence is illustrated by a comparison of the spectrum of poly(rA) · poly(rU) with that of poly(rA–rU) · poly(rA–rU) (Morikawa *et al.,* 1973). The latter polymer has an alternating sequence –AUAUAUAU– on each strand, but forms a double helix with interstrand base pairs in the same way as the homopolymers. Moreover, the geometries of the helices formed from both types of polymer are thought to be the same. At high temperature, both complexes dissociate to give the same spectra. However, the spectra of the two forms of double helix are markedly different, showing that spectral changes resulting from base stacking are dependent on the specific base sequence along each polyribonucleotide strand. Furthermore, the carbonyl profile of poly(rA) · poly(rU) differs from that of poly(rA–rU) · poly(rA–rU). The former has a single C=O feature at 1688 cm^{-1} while the latter yields a peak at 1677 cm^{-1} with a shoulder at 1650 cm^{-1}. Therefore, although the geometries of base pairing are the same for both duplexes, different carbonyl profiles are observed. Vibrational coupling between adjacent base pairs was proposed to explain this finding (Morikawa *et al.,* 1973). Howard *et al.* (1969) invoked vibrational coupling within a G–C base pair to account for some unusual IR spectra of ^{18}O-substituted poly(C) · poly(G). Hence interbase vibrational coupling appears to be an important facet of double-helical structures. In a similar study, Baret and co-workers (1979) investigated the spectral differences of poly(dA) · poly(dT) and poly(dA–dT) · poly(dA–dT).

These considerations require that caution be used when making quantitative statements regarding the degree of base pairing and base stacking in a RNA molecule. Progress toward unraveling the problems of spectral dependence on base sequence and toward separating the spectral consequences of pairing and stacking is being made by using short polynucleotides [e.g., ApG and ApU (Prescott *et al.,* 1974; Gramlich *et al.,* 1978)] as models and by using systems, such as guanosine gels (Savoie *et al.,* 1978; Delabar and Guschlbauer, 1979), which contain well-defined and relatively simple environments for the base.

Inosine is a nucleoside which occurs in small amounts in some RNAs and is structurally similar to guanosine (Fig. 7.1), with guanosine's peripheral $–NH_2$ being replaced by $–H$ in inosine. Brown *et al.* (1972b) proposed that an inosine ring mode at 1464 cm^{-1}, which they assigned to the amide II vibration of the cis amide group, could be used as a quantitative measure of base pairing in poly(rC) · poly(rI). The feature shifts to 1484 cm^{-1} on hydrogen-bonded base pairing.

Although precise interpretation may not be possible, hyper- and hypochromic effects can still provide a wealth of detail regarding the melting of nucleic acids. An example is the work of Erfurth and Peticolas (1975) on the melting of DNA. The Raman spectra of calf-thymus DNA at 25°, 84°, and 95°C are shown in Fig. 7.5. Because the melting point of this sample of DNA is between 80° and 85°C, the spectra reflect changes attributable to melting. By plotting band height as a function of temperature, it was found that certain Raman features from each of the four bases A, T, C, and G undergo a gradual intensity increase prior to the melting point, e.g., the increase in the 1240-cm^{-1} "T" band, throughout the temperature range studied. This is apparent in Fig. 7.5. However, consideration of the features attributable to the deoxyribose-phosphate groups (discussed in the Section II, B) shows that the average conformation of the backbone remains unchanged below the melting point. Therefore, although the backbone conformation of the helix is not disordered by any premelting, the base conformations are, with disruption of the vertical base–base stacking interaction beginning near 50°C and increasing up

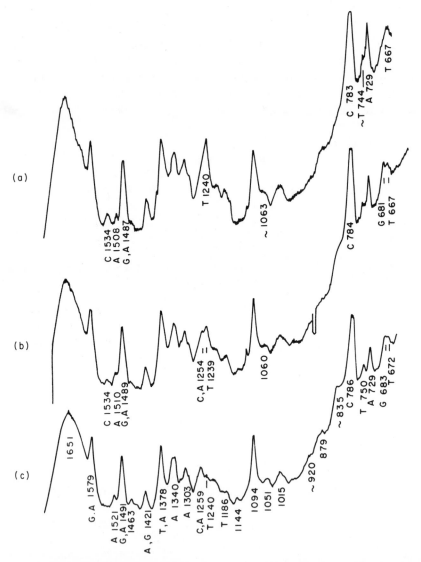

Fig. 7.5. Raman spectra of calf-thymus DNA at (a) 25°, (b) 84°, and (c) 98°C in H_2O, pH 7.0, 0.01 M in sodium cacodylate, and 0.001 M in EDTA. The total Na^+ concentration was 0.075 M. [From Erfurth and Peticolas (1975).]

until the melting point. This experiment illustrates the power of the Raman technique to simultaneously probe different aspects of nucleic acid structure.

B. Conformation of the (Deoxy)ribose–Phosphate Backbone

The phosphate groups

$$
\begin{array}{c}
\text{C} \diagdown \text{O} \quad \text{O} \\
\text{P} \quad \ominus \\
\text{C} \diagup \text{O} \quad \text{O}
\end{array}
$$

in the (deoxy)ribose–phosphate backbone of RNA (and DNA) give rise to two characteristic Raman bands near 800 and 1100 cm^{-1} which are very useful probes of backbone conformation. On the basis of Shimanouchi and co-workers' (1967) analysis, the 800-cm^{-1} feature is assigned to the –O–P–O– symmetrical stretching vibration, and the 1100-cm^{-1} band originates from the PO_2^- symmetrical stretch motion. Conformational utility stems from the fact that the frequency and intensity of the PO_2^- feature are insensitive to backbone geometry, and thus provide an internal standard (Tsuboi *et al.*, 1971), whereas the 800 cm^{-1} feature is highly sensitive to conformation of the –C–O–P–O–C– group and possibly the ribose ring (Erfurth *et al.*, 1972; Brown and Peticolas, 1975).

The Raman spectrum of an ordered, single-stranded or double-helical RNA shows an intense sharp band at 810–815 cm^{-1} which shifts to near 795 cm^{-1} on disordering (Small and Peticolas, 1971; Erfurth *et al.*, 1972; Lafleur *et al.*, 1972). This transition is clearly seen in the spectrum of poly(rA) · poly(rU) in Fig. 7.4, and the same spectrum demonstrates the insensitivity of the 1100-cm^{-1} feature to disruption of secondary structure. The intensity of the PO_2^- Raman mode remains constant at constant ionic strength. For RNA, at low ionic strength, the ratios of intensities of the bands at 815 and 1100 cm^{-1} can be used to monitor the amount of secondary structure. $I_{815} : I_{1100} = 1.66$ in completely ordered ribopolymers (double helical or single stranded) and 0.0 in completely disordered ribopolymers (Thomas, 1975).

The phosphate modes for DNA resemble those for RNA; the DNA 1100-cm^{-1} band is insensitive to changes in secondary structure, and, as in the case of RNA, the features near 800 cm^{-1} in the Raman spectra of DNA are sensitive to conformation. These facets are illustrated by the conformational changes observed in DNA fibers. It was known from X-ray diffraction patterns that oriented fibers of DNA, at the appropriate salt content, underwent conformational transitions with changes in humidity. At 75–92% relative humidity, the DNA is in the so-called A form, but above 92% relative humidity the B form exists. Erfurth *et al.* (1975) were able to follow this transition in the Raman spectrum (Fig. 7.6). The $-O-P-O-$ symmetric stretch of the A form at 807 cm^{-1} shifts to 790 cm^{-1} on conversion to the B form. This shift has been modeled by a normal coordinate analysis (Brown and Peticolas, 1975), and, in addition, can be used to follow the relative amounts of the A and B forms. A weak unassigned band at 835 cm^{-1} is also present in spectra of B-form DNA. A further difference between the spectra of the two forms of DNA occurs in the relative intensities of the 665-cm^{-1} band of guanine and 682-cm^{-1} band of thymine. In the A form, the 665-cm^{-1} band is the stronger, and in the B form the 682-cm^{-1} band is the more intense. The absence of a sharp band near 807 cm^{-1} in any of the spectra of DNA in solution shown in Fig. 7.5 shows that the A form is absent under these conditions. It is probable that the phosphate mode contributes to the cytosine peak at 784 cm^{-1}; this, taken with the broad feature at 835 cm^{-1}, indicates that prior to melting the DNA is in the B form.

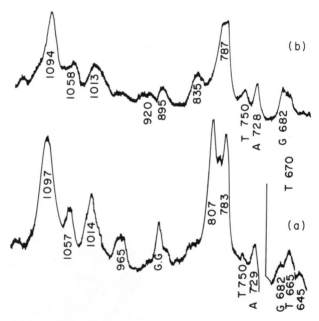

Fig. 7.6. Raman spectra of an oriented fiber of calf-thymus Na-DNA with 3–4% NaCl at (a) 92% and (b) 75% relative humidities. [From Erfurth *et al.* (1975).]

III. Applications to RNA and DNA Structures

A. Transfer RNAs and 5 S RNA

1. Transfer RNA

Transfer ribonucleic acid (tRNA) contains approximately 80 bases in a single strand and is an important component involved in protein synthesis. There is a distinct tRNA molecule for every amino acid, and a protein chain is synthesized by the sequential addition of amino acids carried to the chain by tRNA. The peptide sequence is determined by the fact that as the tRNA presents the amino acid for protein chain elongation it is, at the same time, reading off the genetic code from a messenger RNA molecule. The secondary structure for the tRNA specific for phenylalanine is shown in Fig. 7.2 and, for obvious reasons, is known as the cloverleaf model. Detailed knowledge of the tertiary structure of tRNA has been derived mainly from X-ray diffraction studies of single crystals. However, very few tRNAs have been obtained in crystalline forms that give suitable diffraction patterns. For this reason, as well as to answer questions such as what are the differences between conformers in the crystalline and solution states, the Raman data assume a special importance.

A finding common to all the Raman studies of tRNAs is that, in solution, every tRNA appears to adopt a very similar conformation (Thomas *et al.,* 1972, 1973a; Chen *et al.,* 1978). Based on the relative intensities of the 1100- and 800-cm^{-1} phosphate diester peaks, Thomas and co-workers (1973a) estimated that for *E. coli,* tRNA^{Glu}, tRNA^{Val}, tRNA^{fMet}, and tRNA$_2^{Phe}$ the percentage of nucleotide residues in each which exist in a highly ordered conformation is 84 ± 4%. Similarly, Chen and co-workers (1978) reported that for the 11 native tRNAs examined by them in solution the I_{814}/I_{1100} intensity ratio was 1.73 ± 0.05, indicating a high, relatively constant amount of order in the ribophosphate backbone. These conclusions only hold when salt concentration is carefully controlled since changes in salt concentration can alter the tRNA conformation (Chen *et al.,* 1975; Dobek *et al.,* 1975). Unlike DNA, the backbone conformation of RNA appears to be largely insensitive to gross changes in the degree of hydration. In this regard, Chen and Thomas (1974) showed that the Raman spectra of 16 S and 23 S RNA and tRNA^{Phe} were essentially the same in solution and amorphous powders. Chen and co-workers (1975) went one step further and compared the spectrum of tRNA^{Phe} in solution with that for crystals used for X-ray crystallographic studies. Their finding that the spectra were the same gives confidence in applying the X-ray results to determine the biochemical properties of the functioning aqueous tRNA.

The foregoing results all refer to tRNAs which are not charged, i.e., which do not have their corresponding amino acid covalently bound to them. Thomas and co-workers (1973b) have shown that binding of the amino acid does bring about a conformational change in tRNA. Figure 7.7 compares the spectra of unfractionated (i.e., a mixture of) tRNAs in the uncharged and charged (aminoacylated) states. The amino acids make no observable contribution to the observed spectra, therefore the changes in Fig. 7.7 must result from an alteration of the conformation of the tRNAs by aminoacylation. The intensity increase in the lines at 725 and 785 cm^{-1} on charging with amino acid were ascribed to unstacking of the A and C + U residues, respectively. The small intensity increase of the 814 cm^{-1} phosphodiester feature probably indicates a small increase in backbone order on charging, and the observed changes in the 1150–1450-cm^{-1} region (Fig. 7.7) are also consistent with this conclusion (Thomas *et al.,* 1973b).

2. 5 S RNA

Like tRNA, 5 S RNA and 5.8 S RNA belong to a class of small RNA molecules that function in protein synthesis. At the present time, the structure of 5 S and 5.8 S RNA are not known. The Raman spectrum of *E. coli* 5 S RNA has been reported by Chen and co-workers (1978), and Luoma and Marshall (1978a, b) have investigated *E. coli* 5 S RNA in addition to a eucaryotic 5 S RNA and 5.8 S RNA. Each of these species has a Raman spectrum bearing an overall resemblance to that of tRNA, but there are notable quantitative differences, e.g., the 670-cm^{-1} line is relatively more intense in the 5 S RNAs. Since the 670-cm^{-1} feature is a

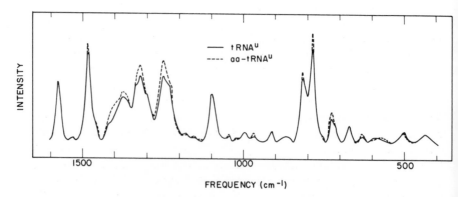

Fig. 7.7. Raman spectra of unfractioned transfer RNA uncharged (tRNA^u) and charged (aa–RNA^u) with amino acid. [From Thomas *et al.* (1973b).]

hyperchromic band attributable to G, it was concluded that, on average, the G bases are more stacked in 5 S RNA compared to tRNA. By using such detailed considerations of the Raman data, taken with other evidence, Luoma and Marshall (1978a, b) were able to propose a model for the secondary structure of 5 S RNA which has an overall resemblance to the cloverleaf model for tRNA shown in Fig. 7.2. However, there are quantitative differences in, for example, the lengths of base-paired regions.

B. DNA

As is the case for many macromolecules, IR studies on DNA preceded the Raman investigations by several years. Reports on the IR spectra of DNA appeared in the 1950s, and this work is reviewed in Parker's (1971) monograph. It was not until 1968 that Tobin (1969) published the first Raman spectrum of DNA. Since that time, Peticolas and co-workers have been particularly active, and some of their major studies on DNA are mentioned in Section II. Additionally, the principles laid down in the earlier studies of the free bases and polynucleotides have been used to study heavy-metal–DNA interactions (Chrisman *et al.,* 1977) and chemical modification of DNA (Mansy and Peticolas, 1976; Herbeck *et al.,* 1976). In the work on heavy-metal binding, the mode of binding of CH_3Hg^+ to DNA was followed as a function of CH_3Hg^+ concentration relative to phosphate. When this ratio is 0.3, extensive binding to both G and T bases is observed. The chemical modification studies of Herbeck *et al.* (1976) consisted of producing cross-linked pyrimidine dimers in double-helical DNA by UV irradiation. These investigators used Raman spectroscopy to demonstrate that circular dichroism provides a reliable measure of conformational change under the conditions employed. Then the circular dichroism measurements were used to monitor the structural consequences of photodimerization and to show that, for example, the pyrimidine–dimer cross-links induced into the B form of DNA lock it irreversibly into that conformation. In another publication, Mansy and Peticolas (1976) were able to follow the consequences of alkylation of DNA by methyl nitrogen mustard (an antitumor drug) and dimethyl nitrogen half mustard. When an excess of alkylating agent was used, the observed Raman frequencies of the G base in DNA changed to those of 7-methylguanosine, showing that essentially all of the guanine bases were alkylated in the N-7 position. No changes were observed in the modes of other bases. Interestingly, in the alkylated double-helical DNA, the 7-methylguanine bases were in the keto form, but on melting, the alkylated guanine changed to the zwitterionic form.

IV. Viruses and Other Nucleic Acid–Protein Complexes

Viruses are, on the biological scale of things, relatively simple entities containing well-defined protein and nucleic acid components. Usually, both components contribute to the observed Raman spectrum of a virus, and by applying the principles stated previously, conclusions are forthcoming on the protein and nucleic acid conformations. Thomas (1976) and co-workers have been particularly active in investigating virus structure by Raman spectroscopy, and we shall discuss two examples from their work. The first concerns bacteriophage MS2, a bacterial virus of particle weight 3.6×10^6. This phage is composed of 180 identical coat-protein molecules each of MW 13,750 which make up the capsid or "coat" of the virus, one additional protein molecule of MW 38,000, and one RNA strand of MW 1.1×10^6. When the virus infects its bacterial host, the single protein is thought to enter the host with the RNA, leaving the capsid protein outside.

Raman spectra of the native MS2 phage in H_2O and D_2O are shown in Fig. 7.8, and the assigned peaks are labeled underneath (Thomas *et al.*,

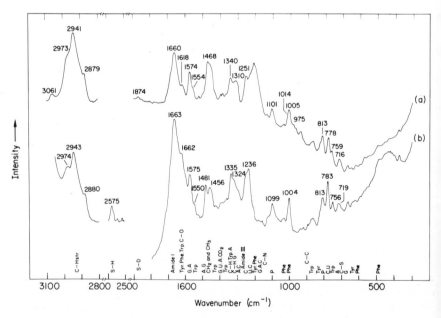

Fig. 7.8. Raman spectra of the native MS2 virus in (a) H_2O, and (b) D_2O solutions. The virus concentration is ~38 mg/ml. Laser excitation is 488.0 nm, 300 mW. [From Thomas *et al.* (1976).]

1976). Equally good quality spectra were obtained for the capsid protein with the RNA removed and of the RNA in the absence of protein. The Raman spectrum of intact MS2 compared to the sum of the spectra of capsid and free RNA were the same, indicating that major conformational changes in protein or RNA did not occur on separation of these components. However, free RNA is more susceptible to changes in secondary structure promoted by increasing the temperature. Thus the viral capsid exerts a significant stabilizing effect on the secondary structure of MS2 RNA. Turning now to a more detailed analysis of Fig. 7.8, the amide III feature at 1236 cm^{-1} indicates a substantial β element in the capsid-protein conformation, roughly estimated at 60 $\pm$ 20%. That the protein conformation is determined mainly by regions of β and random-coil structures is also consistent with the appearance of the amide I band at 1660 cm^{-1}. Another important observation stemming from the coat-protein contribution to Fig. 7.8 concerns cysteine residues of the protein. Since there is no Raman scattering in the 500-cm^{-1} region where ν_{S-S} is expected, but there is evidence for ν_{S-H} at 2574 cm^{-1}, most of the cysteine residues are thought to be present as the reduced S–H form rather than in disulfide linkages.

For the RNA moiety, the intensity of the 813-cm^{-1} line in the intact phage (Fig. 7.8) and RNA alone indicates that in both forms more than 85% of the RNA nucleotide residues participate in base-pairing and/or base-stacking interactions. The temperature dependence of the Raman spectra of protein-free MS2 RNA reveals the following.

1. Loss of hypochromism or hyperchromism, showing that base-stacking interactions of A, U, C, and G are mostly eliminated above 60°C in an essentially cooperative manner.

2. Intensity decay in the phosphodiester band (near 815 cm^{-1}), indicating gradual backbone disordering from 0 to 80°C.

3. Gradual intensity changes in the carbonyl region (1650 cm^{-1}), reflecting a gradual rupture of hydrogen bonds between bases over a wide temperature range.

The second detailed study (Hartman *et al.,* 1978) concerns turnip yellow mosaic virus (TYMV), which is composed of 180 identical coat-protein molecules (each containing 180 amino acid residues) residing in an icosahedral capsid. The capsid contains one molecule of RNA which appears to be in contact with the protein. By using Lippert's method based on the relative Raman intensities in the amide I and III regions (Chapter 4, Section III, B) the secondary structure of the coat protein was calculated to be 9 $\pm$ 5% α-helical, 43 $\pm$ 6% β-sheet, and 48 $\pm$ 6% irregular conformation. The estimate of the β-sheet structure was revised upward from that given in an earlier report by the same group (Turano *et al.*, 1976). Modes attributable to tryptophan, cysteine (–SH), and tyrosine were identified in the Raman spectrum of the virus. The tryptophan and cysteine side chains appear to be exposed to solvent, but only one-third of the available tyrosine residues appears to hydrogen bond to water. The ratio of the ribophosphate 1100- and 815-cm^{-1} features was used to estimate that $77 \pm 5\%$ of the encapsulated RNA was in an ordered form. One proposed mechanism for the stabilization of TYMV at pH $<$ 6 is hydrogen bonding between protonated cytosine residues and –CO$_2$H groups from the protein. In keeping with this mechanism, Hartman *et al.* (1978) were able to find some evidence from the Raman spectrum that at pH 4.8 a considerable fraction of the cytosines in TYMV's RNA was protonated.

In an early study of virus structure (Hartman *et al.*, 1973), a comparison of the spectra of the RNA phage R17 and the protein-free RNA showed that the RNA backbone conformation was, as in the case of MS2, unchanged by complexation to the protein. Similar intensities were found for the ring vibrations with the exception of a band at 1480 cm^{-1} (assigned to guanine and adenine), which is clearly weaker in phage RNA than in protein-free RNA. The intensity difference was ascribed to different molecular environments for guanine. Recently, tobacco mosaic virus (Shie *et al.*, 1978; Fox *et al.*, 1979) and cucumber virus 4 (Shie *et al.*, 1978) have also been characterized by their Raman spectra. Essentially the same type of information, as that discussed for the MS2 and TYMV particles, was forthcoming on protein and RNA secondary structure and protein side chains.

The filamentous bacterial viruses Pf1 and fd differ from the above viruses in that these are DNA–protein complexes. Because of the low DNA content ($\sim$12%), the observed spectra (Thomas and Murphy, 1975) are dominated by features from the coat protein. Spectral differences between the coat proteins of the two viruses reflect different amino acid compositions. Both are found to be highly α helical. In the phosphodiester stretching region, a weak band at 785 cm^{-1} suggests a B- or C-type conformation for the polynucleotide backbone.

The DNA in animal cells is associated with, and may be packaged by, small basic proteins known as histones. Additionally, non-histone proteins (NHP) have been found associated with the DNA. Models for these DNA–protein complexes are of intense interest since these structures are intimately concerned with the ways in which cells replicate and genes are "turned on and off." Based primarily on Raman data, the sites of DNA–protein interactions have been tentatively located in the grooves running along DNA (Goodwin and Brahms, 1978; Goodwin *et al.*, 1979). The principal observations concern the intensities of the 1580- and 1490-cm^{-1}

DNA features in the Raman spectra of various DNA–protein complexes. These complexes were isolated from different animal cells and varied in the amount of non-histone protein they contained. In those complexes rich in NHP, the intensity of the Raman band near 1490 cm^{-1} is markedly decreased with respect to the same band in free DNA. In contrast, for protein–DNA complexes possessing little NHP, the 1490 cm^{-1} feature has the same intensity as that found in free DNA. Because the 1490-cm^{-1} feature is sensitive to interactions at the N-7 position of guanine and because this position is exposed to the large groove in DNA in its B form, it was concluded the NHP binds to the large groove in DNA. This is illustrated in Fig. 7.3.

The Raman spectra also provide a clue to the location of the histone proteins on the DNA. Methylation experiments indicated that the 1580-cm^{-1} DNA band is attributable to a mode involving motion of the N-3 position of the adenine which lies in the small groove of DNA. A reduction in intensity of the 1580-cm^{-1} band was observed for all complexes isolated from different tissues. Because these DNA–protein complexes contained approximately constant amounts of histone proteins, but widely varying amounts of NHP, it was concluded that the histone proteins were perturbing the N-3 position of adenine and therefore must lie in the small groove of DNA. This is also illustrated in Fig. 7.3.

Supporting evidence for the model shown in Fig. 7.3 comes from Raman data on protamine–DNA complexes (Herskovits and Brahms, 1976). Moreover, the different relative intensities of the 1490-cm^{-1} band in protein–DNA complexes from different origins could explain the apparent discrepancy between two previous Raman studies. Mansy et al. (1976) observed a reduction in the intensity of the 1490-cm^{-1} band in the Raman spectra of complexes from mouse myeloma chromatin, whereas Thomas et al. (1977) observed no change in this band in Raman spectra from chicken erythrocyte complexes with respect to DNA or after dissociation of the complex in 2 M NaCl solution.

V. Resonance Raman Studies of Nucleic Acids

A. Using the Absorption Bands of the Natural Bases

The electronic transitions of the nucleic acid bases in the 250–300-nm region are of the greatest importance for quantitative assays and structural studies that use existing optical techniques such as absorption spectroscopy. Thus, the increasing availability of reliable laser sources in the 250–300-nm region opens an important new area since the same absorption bands that have been of value in qualitative structural analysis can now be used to generate RR spectra and the detailed information which may be derived from these. To date, all UV sources have been obtained by frequency-doubling visible laser light. Continuous light at 257.3 nm has been obtained by passing 514.5-nm irradiation through a cooled ammonium dihydrogen phosphate crystal, and tunable UV sources can be derived from frequency-doubling pulsed dye laser output in the visible region. Frequency doubling is an inefficient process, and the resulting average power in the UV is usually of the order of a few milliwatts.

The most detailed analysis of the RR spectra of nucleic acids appears in the review by Nishimura *et al.* (1978). These authors present normal coordinate treatments and RR theory in addition to the experimental results concerning the UV-excited RR spectra of nucleic acids. In this section, only the more important generalizations arising from the research of the Nishimura and co-workers and other groups is outlined.

There is a similarity between the overall appearance of the RR and Raman spectrum of a given nucleotide, although, of course, the resonance spectrum may be intensity enhanced by a factor of 10^5. The similarity of the Raman and RR spectra is illustrated by the spectra shown in Fig. 7.9.

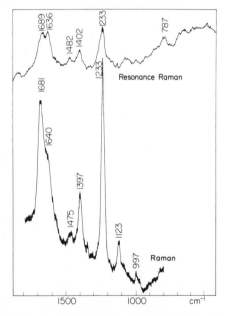

Fig. 7.9. (Top) RR spectrum (257.3-nm excitation) of β-uridine-5′-phosphoric acid [adapted from Nishimura *et al.* (1978)]. (Bottom) Raman spectrum (457.9-nm excitation) of UTP [adapted from Chinsky *et al.* (1978)].

The same uracil ring modes dominate the RR spectrum of β-uridine-5'-phosphoric acid (UMP) and the normal Raman spectrum of uridine-5'-triphosphate (UTP). In the RR spectra of polynucleotides the phosphate modes near 800 and 1100 cm^{-1} are not intensity enhanced and therefore are usually absent. The UV-absorption spectra of nucleic acids are complex and are not completely understood. This is another area in which RR spectroscopy has much to contribute. For the uracil derivatives shown in Fig. 7.9, Nishimura *et al.* (1978) and Chinsky *et al.* (1978) agree that the dominant means of resonance enhancement is derived from the uracil band centered at 260 nm. However, Nishimura and co-workers believe that modes detected at 787 and 560 cm^{-1} gain intensity by coupling to excited states which give rise to absorption bands to lower wavelengths of the 260-nm feature. The complexity of the relationship between the RR intensity and the electronic spectrum is further illustrated by the excitation wavelength dependence of the hypochromicity associated with the 1233-cm^{-1} uracil feature seen in Fig. 7.9. This band undergoes a large change in intensity in the normal Raman spectrum of poly(rU) resulting from an order to disorder transition (Fig. 7.4); the uracil line at 1230 cm^{-1} is less than half as intense at 32°C [in the ordered poly(rA) · poly(rU) duplex] than at 85°C [in the disordered poly(rU) strand]. This Raman hypochromism is, however, a function of excitation wavelength; for poly(rU) it is still evident with 351-nm excitation, but becomes less at 310 and 300 nm until at 295 nm there is no intensity change on formation of disordered poly(rU) (Chinsky and Turpin, 1980). Finally, using 290-nm excitation, Chinsky and Turpin (1980) found evidence for the 1230-cm^{-1} band becoming hyperchromic (i.e., increasing in intensity with increasing order).

Progress toward elucidating the nature of the RR spectrum of adenine has been made by two groups who obtained preresonance or complete excitation profiles. There is a general concensus that adenine has three excited electronic states corresponding to the 276-, 260-, and 210-nm features seen in the absorption spectra. Tsuboi and co-workers (1974), using a preresonance excitation profile, were able to conclude that two bands at 1583 and 1484 cm^{-1} (seen in Figs. 7.4 and 7.10) derive their intensities from the 276-nm absorption band whereas the Raman band at 730 cm^{-1} appears to be associated mainly with the 210-nm absorption. Support for this conclusion came from an experiment on poly(rA–U) · poly(rA–U). The absorption spectrum of this aqueous copolymer shows marked intensity changes on changing temperature from 29° to 81°C. The absorption intensities of the 260- and 210-nm bands both increase by a factor of 1.6 on heating. In contrast, little change is seen in the intensity of the 276-nm feature, although it does shift toward shorter wavelength on heating. The corresponding changes in the 488-nm excited Raman spectra (Morikawa *et al.*, 1973) are that heating increases the intensity of the 730-cm^{-1} adenine band by a factor of 1.9, whereas no change is seen in the 1583- and 1484-cm^{-1} modes. Tsuboi interprets these findings as evidence that the latter modes are not coupled to the 210- or 260-nm absorption bands, but, instead, derive their enhancement for the 276-nm band. Further support for this proposal came from Peticolas's laboratory. Blazej and Peticolas (1977) measured the RR spectra in the 267–305-nm range by using a frequency-doubled pulsed dye laser, and the spectrum of AMP taken using 267-nm excitation is shown in Fig. 7.10. The measured excitation profile was reproduced using an equation from a simple vibronic theory based on A-type intensity enhancement in which the position of E_{qq}, the location of the zero–zero energy level of the *e*th state, and two damping terms were used as variable parameters. The best fit gives the position of the electronic transition responsible for the A-type enhancement of a given mode. The bands at 1484 and 1583 cm^{-1} were shown to gain intensity by coupling with a transition at 276 nm, which is in excellent agreement with the earlier suggestion of Tsuboi *et al.* (1974). The features at 1310, 1338, and 1380 cm^{-1} (Fig. 7.10) were best fitted to their excitation profiles by permitting coupling to an electronic transition near 267 nm. However, for the bands between 1300 and 1400 cm^{-1} some intensity enhancement by a B-type term cannot be ruled out.

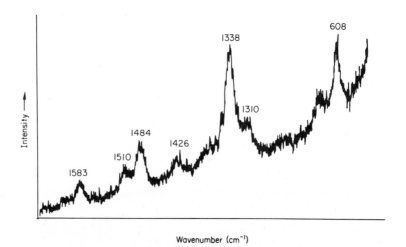

Wavenumber (cm^{-1})

Fig. 7.10. RR spectrum (267-nm excitation) of 0.5 m*M* AMP–0.5 *M* sodium cacodylate, pH 7.1. The band at 608 cm^{-1} is attributable to cacodylate. [From Blazej and Peticolas (1977).]

B. Using the Absorption Bands of Modified Bases–Resonance Raman Labels

By placing a bromine atom on the periphery of a base, the absorption spectrum shifts toward the red, and there is significant absorption at 300 nm. Excitation by 300-nm irradiation gives rise to a RR spectrum of the bromo derivative, and if the bromo derivative occurs in a nucleic acid along with other unmodified bases, the RR spectrum of the brominated base alone will be obtained. Thus derivatives such as 8-Br–ATP, 8-Br–adenosine, 8-Br–guanosine, 5-Br–cytidine, and 5-Br–deoxyuridine (Chinsky *et al.*, 1978) can be considered to be RR labels. The last compound, 5-Br–deoxyuridine, is particularly important because it has been widely used in biological investigations of DNA replication and transcription. Figure 7.11 illustrates the finding of Chinsky and co-workers (1977) that modes attributable to 5-Br–deoxyuridine dominate the RR spectrum of the deoxyribopolymer containing 5-Br–deoxyuridine and adenine. In melting studies, the bands at 1627, 1352, and 1220 cm^{-1} were found to be markedly hyperchromic. The predominance of the uracil features in Fig. 7.11 contrasts with the 257.3-nm excited RR spectrum of poly(rU) · poly(rA) in which only adenine features were observed (Pézolet *et al.*, 1975).

Sulfur-substituted bases also have red-shifted absorbances which can be used to obtain RR spectra by excitation in the 300–360-nm range. These bases could be used as RR labels, but they also occur occasionally as natural substituents. Nishimura and co-workers (1976) obtained RR spectra of the single 4-thiouridine residue which occurs in several tRNAs. Figure 7.12 shows the 363.8-nm excited RR spectra of tRNA^{fMet} and tRNA^{Gly}. Surprisingly none of the lines in Fig. 7.12 correspond to the peaks in the RR spectrum of aqueous 4-thiouridine. However, when the tRNA spectra were obtained using a spinning Raman cell the peaks seen in Fig. 7.12 were replaced by those of 4-thiouridine. This and other data enabled Nishimura *et al.* (1976) to conclude that, in the stationary sample, the 4-thiouridine within the tRNA was photochemically modified by the laser beam, probably by cross-linking to a proximal cytidine residue.

The RR-labeling approach has also been used to study the interaction between chromophoric drugs and DNA; hence in these cases vibrational spectra are obtained on the drug and features attributable to the DNA are not observed. The RR spectrum of the chromophoric portion of adriamycin, an anthracycline antitumor antibiotic, in the free state (Hillig and Morris, 1976) and bound to DNA (Manfait *et al.*, 1980) have been reported. The intensity changes seen on binding were interpreted in terms of strong interactions about the chromophore's –OH groups. Intensity

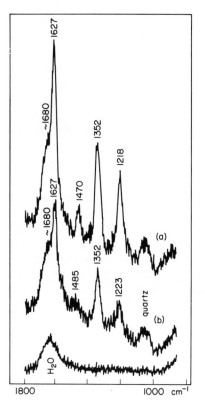

Fig. 7.11. Comparison of RR spectra of (a) the monomer 5-Br–deoxyuridine and (b) the polymer polyd(BrU–A) measured under identical condition: base concentration 1.3×10^{-4} M and 300-nm excitation wavelength. [From Chinsky *et al.* (1977).]

changes were also observed in the RR spectrum of actinomycin D on binding to DNA (Chinsky *et al.*, 1975). Actinomycin's phenoxazone chromophore gave rise to intense RR bands at 1505, 1489, 1405, 1385, and 1265 cm⁻¹ when excited with 457.9-nm light. Chinsky *et al.* (1975) found that the intensity ratio of the 1489- and 1505-cm⁻¹ peaks provides an estimate of binding affinity and could possibly provide a probe of the drug–DNA interaction at the molecular level. Another drug–DNA complex recently studied by Raman spectroscopy involves the oligopeptide antibiotics netropsin and disamycin A (Martin *et al.*, 1978). These antibiotics do not have an electron-absorption band in the visible spectrum; consequently, the normal Raman spectra of the complexes, containing both drug and DNA features, were obtained. The spectrum of the DNA was subsequently subtracted using computer methods. The several changes

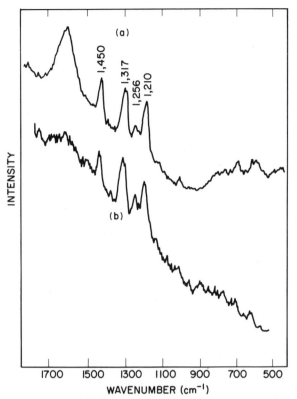

Fig. 7.12. RR spectra (363.8-nm excitation) of (a) tRNA[fMet] and (b) tRNA[Gly]; both from *E. coli.* All of the observed RR features can be attributed to photochemically modified 4-thiouridine. (Reprinted by permission from *Nature,* Vol. 260, No. 5547, pp. 173–174. Copyright © 1976 Macmillan Journals Limited.)

observed in the drugs' Raman spectra on binding to DNA were consistent with a model in which the drugs' N–H groups hydrogen bond to the nucleic acid.

Suggested Reviews for Further Reading

Nishimura, Y., Hirakawa, A. Y., and Tsuboi, M. (1978). Resonance Raman spectroscopy of nucleic acids. *Adv. Infrared Raman Spectrosc.* **5**, 217.

Peticolas, W. L. (1975). Applications of Raman spectroscopy to biological macromolecules. *Biochimie* **57**, 417.

Thomas, G. J. (1975). Raman spectroscopy of biopolymers. *Vib. Spectra Struct.* **3**, 239.

Thomas, G. J., and Kyogoku, Y. (1977). Biological science. *In* "Infrared and Raman Spectroscopy," Part C (E. G. Brame and J. G. Grasselli, eds.), Vol. 1, p. 717. Dekker, New York. Both this and the previous review contain comments on nucleic acids and other biological molecules.

CHAPTER 8

Lipids, Membranes, and Carbohydrates

I. Structures of Lipids and Membranes

Membranes are organized assemblies consisting mainly of proteins and lipids. Membranes separate the "inside" from the "outside" of the living cell and, at the same time, provide means of transporting certain materials across the membrane while posing an impermeable barrier to other molecules. Some membranes contain receptors that recognize external stimuli. For example, membranes involved in the visual process respond to photons. In other systems, bacteria can migrate toward food in response to specific "food receptors" on the bacterial cell membrane. In these cases the stimuli control cell function, and thus membranes play an important role in biological communication. Energy conversion is another process in which membranes play a major role. Both photosynthesis, in which light is converted to chemical energy, and oxidative phosphorylation, in which adenosine triphosphate is formed as a useful energy source, take place in different membrane assemblies.

Membranes are sheetlike structures, usually from 60 to 100 Å in thickness, in which the lipid and protein components are held together by noncovalent forces. The protein components vary widely from membrane to membrane, and specific proteins mediate the distinctive function of membranes. Proteins can function as receptors, transmembrane pumps and gates, energy convertors, and enzymes. However, the lipid compo-

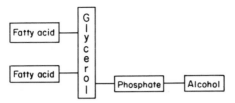

Fig. 8.1. Representation of a phosphoglyceride, a phospholipid based on glycerol.

nents of membranes are less diverse and easier to characterize. In this chapter we outline the chemical structures of phospholipids, which are the most widely occurring lipid component in membranes and which, more-over, have been the most widely studied by Raman spectroscopy.

A. Phospholipid Structures

Phospholipids are the major class of lipids found in membranes. Phospholipids are usually derived from glycerol and consist of a glycerol backbone, two fatty acid chains, and a phosphorylated alcohol (see Fig. 8.1).

The fatty acid chains in phospholipids usually contain an even number of carbon atoms between C_{14} and C_{24}. In phosphoglycerides the C-1 and C-2 hydroxyl groups of glycerol are esterified to the carboxyl groups of two fatty acid chains. The C-3 hydroxyl group of the glycerol backbone is esterified to phosphoric acid. In the major phosphoglycerides the phosphate group is further esterified to the hydroxyl group of one of several alcohols. The most frequently occurring alcohols are serine, choline, ethanolamine, glycerol, and inositol. A phosphatidyl choline is shown in Fig. 8.2. The structures of the other principal phosphoglycerides, e.g., phosphatidyl serine, phosphatidyl ethanolamine, phosphatidyl inositol, and diphosphatidyl glycerol, are obtained by replacing the alcohol portion

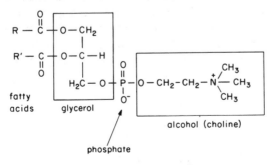

Fig. 8.2. Phosphatidyl choline.

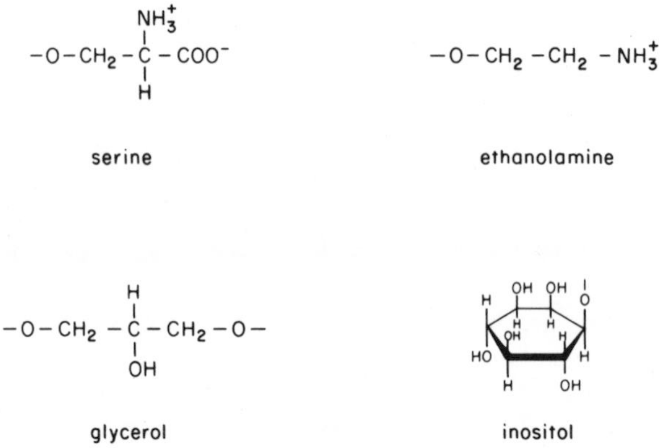

$$-O-CH_2-\underset{\underset{H}{|}}{\overset{\overset{NH_3^+}{|}}{C}}-COO^-$$

serine

$$-O-CH_2-CH_2-NH_3^+$$

ethanolamine

$$-O-CH_2-\underset{\underset{OH}{|}}{\overset{\overset{H}{|}}{C}}-CH_2-O-$$

glycerol

inositol

Fig. 8.3. Common alcohol head groups for phosphoglycerides.

in Fig. 8.2 with one of the constituents shown in Fig. 8.3. In the case of diphosphatidyl glycerol two units of phosphate–glycerol–fatty acid are attached, one to each of the primary alcohols.

A structural theme common to membrane lipids is their amphipathic character; i.e., they contain a polar, hydrophilic head group but the long fatty acid "tails" are nonpolar and hydrophobic. Schematically, a lipid may be represented as shown in Fig. 8.4.

In a mixture of lipid and water molecules, one possible association satisfying both the hydrophilic and hydrophobic properties of membranes lipids is the lipid bilayer shown in Fig. 8.5. In fact, the lipid bilayer is the favored structure for most phospholipids in aqueous media. By mechanically agitating an appropriate lipid in excess water, stable multilamellar particles, known as liposomes, are formed. In these, each of the rings of the onionlike structure consists of a bilayer. In turn, the multilamellar particles may be reduced to single-shell vesicles ($\sim$275 Å in diameter) by ultrasonic waves. Both liposomes and vesicles undergo the characteristic phase transition depicted in Fig. 8.6. In the *gel phase* the hydrocarbon chains are essentially in the all-trans conformation about the C–C bonds, as shown by the straight lines in Fig. 8.6. However, in the liquid-crystalline phase, the hydrocarbon chains possess a population of gauche

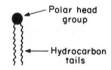

Fig. 8.4. Representation of a phospholipid molecule.

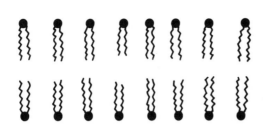

Fig. 8.5. Diagram of a side-on view of a lipid bilayer formed by phospholipid molecules.

C–C isomers (shown as wiggly lines in the figure). The consequent changes in bilayer geometry are also recorded in Fig. 8.6. Compared to the gel phase the liquid-crystalline phase brings about a reduction in bilayer thickness and an increase in the average lateral distance between head groups.

B. A Simple Model for a Membrane

A model for a membrane structure is depicted in Fig. 8.7. It consists of a lipid bilayer which is associated with various kinds of proteins. In the section shown in Fig. 8.7 the protein is either bound to the outside of the layer, in which case it is referred to as an extrinsic (alternatively called peripheral) protein, or the protein is embedded in the phospholipids in

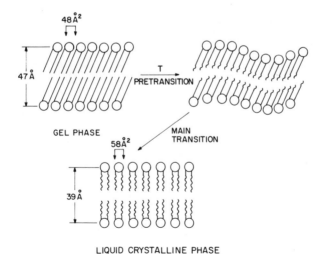

Fig. 8.6. Representation of the gel-to-liquid-crystalline phase transition. (Reproduced from Levin, 1980, by permission.)

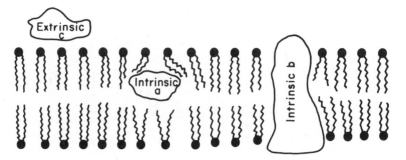

Fig. 8.7. A model for a membrane containing a phospholipid bilayer and intrinsic (a, b) and extrinsic (c) proteins.

some fashion, in which case it is referred to as an intrinsic or integral protein.

It should be stated that although the chemical structures of the lipid components of membranes are precisely documented, the structures of membranes themselves are not as thoroughly understood. Therefore, any statements on the structure of membranes should be regarded as proposed models which may require further modification and may apply to certain membranes but not to others.

II. Raman Spectra of Lipids and Membranes

Raman spectroscopic studies of lipids and membranes offer several advantages over other physical techniques used to characterize these systems. The Raman spectrum provides an instantaneous "snapshot" on the time scale of a fraction of a picosecond. Hence the line broadening which occurs in magnetic resonance spectra, attributable to relaxation phenomena, is absent in the Raman effect. Moreover, the Raman method does not require a probe molecule, and it has the further advantage of being able to monitor both gel and liquid-crystal hydrocarbon regions. The application of the technique to biomembranes has evolved from studies of simple lipids and model membranes which, in turn, grew out of the pioneering studies of aliphatic hydrocarbons by several vibrational spectroscopists. Although the Raman technique has dominated the field in the past decade, it is becoming apparent that Fourier transform infrared spectroscopy will also make future contributions to our knowledge of the vibrational properties of lipids and membrane systems (Cameron *et al.*, 1980; Casal *et al.*, 1980).

A. Simple Lipids and Model-Membrane Systems

The Raman spectra of phospholipids predominantly contain bands arising from vibrations of the hydrocarbon chains, with some contributions from those of the polar head group. Features from the head-group vibrations can be identified, and detailed assignments have been made for the glycerophosphorylethanolamine (Akutsu and Kyogoku, 1975) and glycerophosphorylcholine (Koyama *et al.,* 1977) head groups. The latter study showed that head-group conformations are unchanged at temperatures where the hydrocarbon chains begin to melt. Thus the Raman features attributable to certain head-group modes can be used as temperature invariant internal standards, a finding which is often used in studies of the hydrocarbon chain conformations described below.

Analysis of the Raman spectra attributable to the fatty acid chains is facilitated by the extensive work available on the vibrational spectra of paraffins (Schachtschneider and Snyder, 1963; Snyder and Schachtscheider, 1963; Schaufele and Shimanouchi, 1967; Snyder, 1967; Schaufele, 1968). In an early laser-Raman study, Schaufele (1968) demonstrated that the Raman spectra of a number of straight-chain hydrocarbons were sensitive to a change of phase. Schaufele's report concerned polymethylenes with the general formula $CH_3(CH_2)_nCH_3$, with $n = 5-36$. Other investigators (Tasumi *et al.,* 1962; Schaufele and Shimanouchi, 1967) studied the spectra of polyethylene, which is essentially an infinite chain consisting of a very large number of methylene groups. As a consequence of this total effort the vibrational motions of long hydrocarbon chains in the straight, all-trans conformation are well understood. In a highly ordered linear polymer, such as polyethylene, the frequency ascribable to the motion of an individual CH_2 group has little conceptual value. Instead, one must consider the motion of the entire chain; that is, the normal modes involve the complete chain and are not localized within a few atoms.

Before we consider the behavior of a long $(-CH_2-)_n$ chain, we first introduce some of the concepts inherent in the vibrations of polymers by using two mechanical models. The approach used is similar to that discussed by Peticolas (1979). Figure 8.8 consists of a "balls and springs" model of a linear polymer. Each little oscillator is caused to vibrate at right angles to the long axis of the "polymer." When the oscillators start at the same equilibrium position and vibrate in phase, the wavelength of the vibration is the same as the length of the polymer. Therefore if the polymer is effectively infinite in length, then the wavelength λ is infinite. This motion with phase angle $\theta = 0$ is shown in Fig. 8.8b. At the other

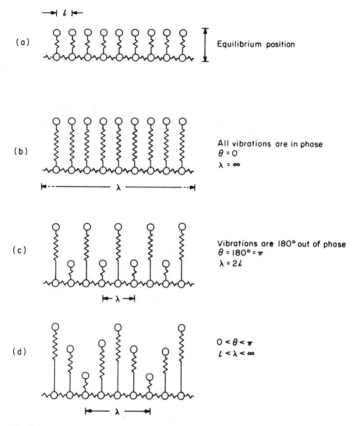

Fig. 8.8. Diagrams of various transverse optical phonons in a chain molecule. [From Peticolas (1979).]

extreme, the alternate oscillators may be completely out of phase. This situation is depicted in the Fig. 8.8c. As one oscillator reaches its maximum amplitude, its immediate neighbors are at their minimum amplitudes. Between the extremes of in-phase and out-of-phase oscillations there is the general sine-wave motion, shown in Fig. 8.8d, in which each oscillator is out of phase with its neighbor by an amount θ and $0 < \theta < \pi$. These wavelike motions have a characteristic wavelength λ, and occur perpendicular to the chain axis. They are known as *transverse phonons* since the quantized waves within polymers are often called *phonons*.

Collective motions may also occur along the long axis of the polymer. Important modes of this type are the *longitudinal acoustical modes* (LAMs), which involve stretching and compression of the polymer chain.

A longitudinal phonon is shown diagrammatically in Fig. 8.9. The longest wavelength mode has a wavelength twice the length of the polymer; the shortest wavelength mode is $2l$, where l is the equilibrium distance between the chemical repeat units (the balls in Fig. 8.9). In a general treatment, it can be shown that for a linear polymer with m atoms per translational repeat unit (e.g., $m = 6$ for the $-C_2H_4-$ translational repeat group in a polyethylene chain) and n repeat units, there are $3mn$ vibrations which occur in $3m$ frequency branches. Within each branch the frequency of each of the n possible modes depends on the relative phase of the atomic displacements in the other repeat units. Each mode within a branch is characterized by a phase angle

$$\theta = k\pi/n, \qquad k = 0, 1, 2, \ldots, n - 1 \qquad (8.1)$$

and the frequencies within each branch fall within a rather narrow range. The curves of frequency versus phase angle for polyethylene are shown in Fig. 8.10. On each of these curves lie n frequencies which form a smooth curve between $\theta = 0$ and π. The dispersion curves fall into two classes; in one class the atoms within the chemical units vibrate against each other. This is exemplified by the model shown in Fig. 8.8, and the resultant modes are known as *optical phonons*. In *acoustical phonons*, the repeat units themselves are displaced with respect to each other, and this is exemplified by the longitudinal motion shown in Fig. 8.9. As the wavelength of the acoustical phonons increases, the motions correspond to the breathing and bending motions of the entire polymer.

For an infinitely long chain (e.g., polyethylene) only those modes with $k = 0$ (i.e., where the motions of corresponding atoms in each repeat unit are in phase) are allowed in the Raman spectrum. For finite lengths of chain, modes with other k values may also occur in the Raman spectrum. In a situation where a conformational change in a $(-CH_2-)_n$ chain from all-trans to non-all-trans occurs, the useful Raman bands for monitoring this change will be those within a branch for which the frequency $k = 0$ is different from that for $k \neq 0$. This occurs because as the conformational change takes place, the band attributable to $k = 0$ mode will decrease in

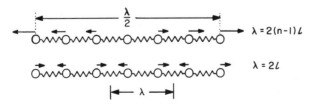

Fig. 8.9. Diagram of a longitudinal optical phonon. [From Peticolas (1979).]

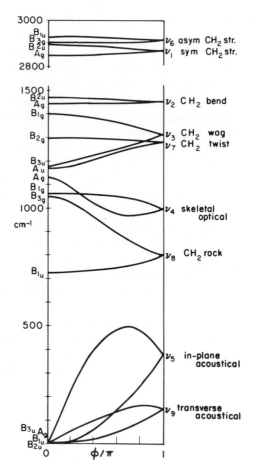

Fig. 8.10. Phonon dispersion curve of a polyethylene zigzag chain. [From Lippert and Peticolas (1972).]

Raman intensity and will, in principle, appear elsewhere in the spectrum at the position appropriate for $k = 1, 2$, etc. In practice the two regions in which the LAM and skeletal optical modes occur (ν_5 and ν_4 in Fig. 8.10) are useful for studying polymethylene backbone conformation by Raman spectroscopy.

1. Accordionlike (LAM) modes below 600 cm^{-1}

For short-chain crystalline hydrocarbons the region below 600 cm^{-1} in the Raman spectrum shows a series of bands, the wavenumbers of which vary inversely with chain length. These bands result from longitudinal acoustical modes which change the shape of the zigzag polymer in an accordionlike fashion (Fig. 8.11). Schaufele and Shimanouchi (1967) fitted the positions of the LAM bands to a power series expansion in terms of the number of CH_2 units in the chain. Using this relationship between peak position and chain length, Warren and Hooper (1973) were able to analyze mixtures of fatty acids in the C_{12}–C_{24} range. Okabayashi and co-workers (1974) have analyzed the LAM modes of surfactants consisting of sodium alkyl sulfates and potassium aliphatic carboxylates. Changes between spectra of the surfactants on going from the aqueous to the solid phase indicated the presence of gauche isomers in solution for those fatty chains containing more than eight carbon atoms. Scherer and Snyder (1980) were able to calculate the energy difference ΔE between the gauche and trans forms of the C–C bond by measuring the temperature dependence of the LAM-like modes of the $n = 11$–14 n-alkanes. They determined a value of ΔE of 508 ± 50 cal/mole.

A Raman study (Brown *et al.*, 1973) of a sonicated dispersion of dipalmitoyl phosphatidyl ethanolamine was consonant with the findings for simpler hydrocarbons. Below the melting temperature T_m, a sharp band at 161 cm^{-1} was observed from the dispersion, at exactly the same wavenumber as from palmitic acid. However, above T_m the 161-cm^{-1} band broadens and moves to higher wavenumbers, as expected from the shorter effective chain lengths resulting from the many possible gauche structures formed at higher temperatures. One difficulty in using the low-wavenumber region for conformational studies arises from the high intensity of scattered light from turbid lipid or membrane solutions. The scattered light can give rise to a very high spectral background near the Rayleigh frequency and obscure the Raman bands at small wavenumber shifts.

2. The 1000–1150-cm^{-1} region

The region between 1000 and 1150 cm^{-1} contains vibrations in which alternate carbon atoms move in opposite directions along the chain length. There are at least three bands which exhibit marked changes on disorder-

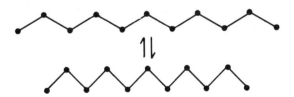

Fig. 8.11. The longitudinal acoustical "accordion" motion of an all-trans hydrocarbon chain.

ing of the hydrocarbon chain. The first report of spectral changes in this region for a phospholipid was by Lippert and Peticolas (1971), who sonicated a suspension of dipalmitoyl lecithin (lecithin is the trivial name for phosphatidyl choline). Their spectra, recorded as the suspension passes through the gel to liquid–crystal transition, are shown in Fig. 8.12. The intensities of the two strong lines at 1128 and 1064 cm^{-1}, assigned to skeletal optical modes of all-trans conformers, decrease abruptly as the melting temperature ($T_m = 39°C$) is approached. In the liquid-crystal phase, above T_m, a line appears at 1089 cm^{-1} which is ascribed to C–C vibrations of random liquidlike conformations. The 1089-cm^{-1} band merges with the band appearing at 1100 cm^{-1} in the low-temperature spectra. The latter is assigned to a stretching vibration of the PO_2^- groups (Lippert and Peticolas, 1972; Mendelsohn *et al.*, 1975a; Spiker and Levin, 1975), and this PO_2^- mode also moves to lower wavenumbers during the transition through T_m. A plot of intensity of the 1128-cm^{-1} band (normalized with respect to a temperature invariant band) versus temperature reveals that the transition seen in Fig. 8.12 is highly cooperative. The effect of perturbants on the shape of these curves has been widely studied and is discussed in Sections II,A,6 and II,B.

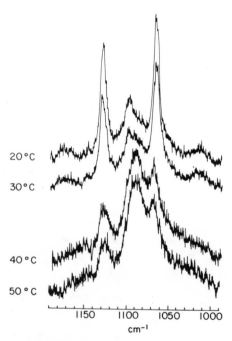

Fig. 8.12. Raman spectra of the 1100-cm^{-1} region of 20% (by weight) DL-dipalmitoyl lecithin sonicates in water as a function of temperature. [From Lippert and Peticolas (1971).]

The temperature dependence of the spectra, in the 1000–1200-cm^{-1} region, of a variety of phospholipids has been studied by a number of investigators. The observed intensity changes for dipalmitoyl lecithin (Brown *et al.*, 1973; Lippert and Peticolas, 1971; Mendelsohn *et al.*, 1976a), dioleyl lecithin (Lippert and Peticolas, 1972), dipalmitoylphosphatidylethanolamine (Brown *et al.*, 1973), dilaurylphosphatidylethanolamine (Mendelsohn *et al.*, 1975a), bilayers of lecithin (Faiman and Long, 1975), and various phosphatidylcholines in water (Spiker and Levin, 1976a; Yellin and Levin, 1977a, b) are qualitatively similar to those shown in Fig. 8.12 for dipalmitoyl lecithin.

There have been several interesting attempts at using the feature near 1130 cm^{-1} to extract quantitative information regarding lipid behavior. Gaber and Peticolas (1977) proposed that the mode near 1130 cm^{-1} was attributable to the sum of intensities from individual all-trans chain segments. The Raman intensity I at 1133 cm^{-1} is thus related to the probability P_n of an all-trans chain containing n C–C bonds by

$$I = I_0 \sum_{n=3}^{N} n C_n P_n \qquad (8.2)$$

where I_0 is the intensity per "trans-unit," n is the number of bonds in an all-trans sequence within a molecule of N bonds, and C_n is the number of times the nth sequence reoccurs. Gaber and Peticolas (1977) then went on to develop an order parameter, S_{trans}, by which the order within different phospholipid chains could be referred to dipalmitoyl phosphatidylcholine (DPPC) as a standard;

$$S_{\text{trans}} = \frac{(I_{1133}/I_{\text{ref}}) \text{ observed}}{(I_{1133}/I_{\text{ref}}) \text{ solid DPPC}} \qquad (8.3)$$

I_{ref} may be either the band intensity at 1090 cm^{-1}, or, for accurate work, a head-group vibration at 722 cm^{-1}. The normalization with respect to DPPC assumes that in the solid form of the latter all the CH_2 groups are in the trans form. Simple relationships of the kind expressed in Eq. (8.3) are of value in semiquantitative analysis. However, their value for exact analysis has been called into question by Karvaly and Loshchilova (1977) and more recently by Pink *et al.* (1980). The latter point out, on the basis of Snyder's (1967) work, that some non-trans isomers may contribute to the 1130-cm^{-1} band. Moreover, Pink *et al.* (1980) concluded from the well-documented gradual decrease of the 1130-cm^{-1} feature as T_m is approached for bilayers of DPPC that conformers other than the all-trans state occur in the hydrocarbon chains at low temperatures. Pink and co-workers (1980) take these factors into account in a detailed theoretical description of the 1130-cm^{-1} band. First, they consider the number and type of conformers likely to be present in the hydrocarbon chains, then they calculate the Raman intensity expected at 1130 cm^{-1} for each state of the chain. In their intensity calculations they rely on Snyder's (1967) normal coordinate analysis of hydrocarbon chains and Theimer's (1957) classical model for calculating Raman intensities from polymer chains. The success of this approach may be judged by considering Fig. 8.13, which compares experiment and theory for DPPC. The gradual experimental diminution of the 1130-cm^{-1} peak with increase in temperature is accounted for by the theory, and there is satisfactory agreement over the entire temperature range. Pink and co-workers comment that since their work demonstrates that there is no simple relationship between the number of trans bonds in a chain and the relative Raman intensity at 1130 cm^{-1}, order parameters of the kind expressed in Eq. (8.3) should be considered qualitative, and attempts to derive accurate enthalpy differences between rotational isomers by using the 1130-cm^{-1} band (Yellin and Levin, 1977a, b) may not be reliable.

3. The C–H stretching region between 2800 and 3000 cm^{-1}

The spectral range $2800–3000 \text{ cm}^{-1}$ contains C–H stretching modes from the methylene and methyl groups of phospholipids. The relative abundance of the methylene groups results in $-CH_2-$ modes dominating the C–H stretching region. This is evidenced by the Raman spectrum,

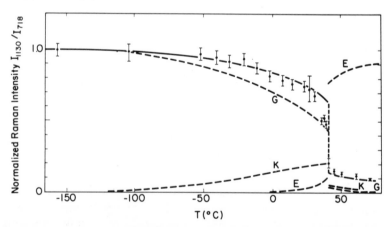

Fig. 8.13. Values of the normalized Raman intensity I_{1130}/I_{718} for DPPC bilayers, as a function of temperature. The 718-cm^{-1} peak is a temperature-invariant feature. Points with error bars are the measured values; the solid line is the theoretically derived curve. Also, the probabilities for a hydrocarbon chain to be in its all-trans state (G), any kink state (K), or its excited melted state (E) are shown as dashed lines. (Reprinted with permission from D. A. Pink *et al.*, *Biochemistry* **19**, 349 (1980). Copyright 1980 American Chemical Society.)

between 2800 and 3000 cm^{-1}, of 1,2-dipalmitoyl-DL-lecithin shown in Fig. 8.14, where the intense peaks at 2847 and 2883 cm^{-1} are assigned to the methylene symmetric and asymmetric stretching modes, respectively (Bulkin and Kirshnan, 1971; Faiman and Larsson, 1976; Verma and Wallach, 1977). The weaker bands at 2962 and 2936 cm^{-1} are assigned to the terminal CH$_3$ asymmetric and symmetric stretches, respectively. The latter assignments are supported by the work of Sunder *et al.* (1976) on stearic acid and its 18-d_3 derivative in the solid state. Although these assignments establish the origin of the "sharp" features seen in the C–H stretching region, there has been considerable effort recently to elucidate the origins of the broad band or bands underlying the resolved peaks. An understanding of these broad bands is essential to the correlation of spectral with conformational and environmental change.

The sensitivity of the C–H region to conformational change was recognized in some of the pioneering studies in the application of Raman spectroscopy to lipid structure (Bulkin and Krishnamachari, 1972; Larsson, 1973; Larsson and Rand, 1973; Bulkin, 1976). Whereas the C–C stretching reflects conformation within the individual fatty acid chains, it was soon realized that the C–H region was also sensitive to interactions between chains (Larsson, 1973). The effort to disentangle inter- and intrachain effects on the Raman spectrum are, of course, related to the problem of assignment. One contribution to the understanding of assignment was made by Bunow and Levin (1977a), who showed that the broad,

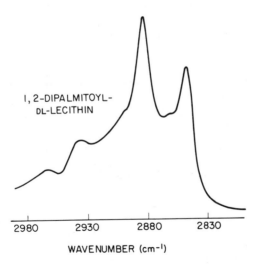

Fig. 8.14. A Raman spectrum of polycrystalline 1,2-dipalmitoyl-DL-lecithin in the C–H stretching region. [From Spiker and Levin (1975).]

structure-sensitive feature which appears at 2930 cm^{-1} in phospholipid liquid-crystal phases originates from asymmetric C–H stretching vibrations of $-CH_2-$ groups that are inactive in the Raman spectrum of the all-trans conformer. However, most attention has been focused on the height ratio of the peaks at 2885 and 2850 cm^{-1} (I_{2885}/I_{2850}) as a monitor of inter- and intrachain forces. Gaber and Peticolas (1977) realized that this ratio is affected by both chain conformation and chain packing, and they proposed that "lateral crystal-like order" could be defined semiquantitatively in terms of (I_{2885}/I_{2850}). Subsequently, Snyder and co-workers (1978, 1980) have been able to provide considerable insight into the factors affecting (I_{2885}/I_{2850}). Broad features underlying the asymmetric C–H stretching mode at 2885 cm^{-1} are responsible for the sensitivity of its intensity to chain packing. The origin of the broad band is a Fermi resonance between the symmetric methylene C–H stretch and various binary combinations of methylene bending modes near 1450 cm^{-1} (Snyder *et al.*, 1978). In a further study, Snyder *et al.* (1980) attempted to define the meaning of lateral crystal-like order by comparing the temperature dependence of the Raman spectra of crystalline n-$C_{16}H_{34}$ and the urea clathrate of n-$C_{16}H_{34}$. Based on these observations, they showed that (I_{2885}/I_{2850}) is affected by two different factors, chain packing and chain mobility. The latter is associated with the freedom of an extended chain to rotate and twist about its long axis. Hence Snyder *et al.* (1980) concluded that (I_{2885}/I_{2850}) should not, by itself, be presumed to yield a unique quantitative measure of lateral order in biomembranes.

A comparison of the C–H and C–C stretching regions provides insight into the changes occurring in phospholipid structures on sonication. The technique of sonication, utilizing ultrasonic sound, is used to transform large heterogeneous phospholipid aggregates into smaller vesicles. These vesicles are spherical and often consist of a single bilayer; they are used as model systems to study processes such as membrane transport. For egg lecithin and dipalmitoyl lecithin, Raman studies (Mendelsohn *et al.*, 1976a; Spiker and Levin, 1976b) showed that gentle bath sonication resulted in changes in the C–H stretching region but little or no perturbation in the C–C stretching region. Thus sonication brings about a change in lateral chain–chain interactions, as a result of the small radius of curvature in the vesicles, but no change in the number of gauche isomers. However, Gaber and Peticolas (1977), using probe sonication, did note a change in the number of trans units on forming vesicles. Thus significant changes may be brought about in the trans/gauche ratio by sonication, depending on the method of preparation. Invariably, however, sonication leads to a perturbation of chain–chain interactions.

Gaber *et al.* (1978a) used Raman difference spectroscopy to study the conformation of dipalmitoyl lecithin in three well-defined temperature ranges. On the basis of peaks observed near 2860 cm^{-1}, they suggested that below the pretransition temperature the triclinic crystal structure exists. Between the pretransition and the main melting temperature T_m 1–2 gauche rotamers per chain are formed with resultant disruption of chain–chain interactions. Moreover, it was proposed that the gauche isomers occur only at the end of all trans oligomers. Above T_m many gauche isomers are formed.

4. Using deuterated lipids as specific and selective probes

The examples discussed in the previous section demonstrate that the C–H stretching region contains a wealth of information regarding lipid conformation and environment. However, in multicomponent systems difficulties arise from the overlapping of C–H features because of chemically different lipids, or in lipid–protein arrays from protein C–H stretching bands. In these systems, it becomes impossible to monitor a single lipid component. Mendelsohn *et al.* (1976b) showed how this problem can be overcome by the use of deuterated components. Mendelsohn *et al.* (1976b) inserted a completely deuterated fatty acid into a model membrane system and followed the C–D stretching vibrations in the spectral window 2000–2220 cm^{-1} which is uncluttered by modes from other components. As the membrane passed through a gel–liquid-crystal transition the linewidth of the C–D stretching vibrations of the bound fatty acid was found to be a sensitive probe of membrane polymethylene chain order. Figure 8.15 compares the temperature variation of the 2103-cm^{-1} C–D mode in pure stearic acid-d_{35} and for the deuterated acid in lecithin multilayers. The melting temperature measured for the bound acid is very similar to that measured directly by taking I_{2880}/I_{2850} for the lecithin.

The use of a deuterated phospholipid in a multicomponent system was reported by Mendelsohn and Maisano (1978), who studied a mixture of dimyristoyl phosphatidylcholine (and its -d_{54} derivative) and distearoyl phosphatidylcholine. Two distinct melting regions were observed for a 1 : 1 molar ratio mixture. The use of the deuterated component allowed the lower (~22°C) transition to be identified primarily with the melting of the shorter chain component and the higher (~47°C) transition to be identified mainly with the melting of the longer chains. Interestingly, the C–H stretching vibrations of the distearoyl component responded to the melting of the dimyristoyl component. Further experiments along these lines have been reported for dipalymitoylphosphalidylcholine–dipalymitoyl-phosphatidylethanolamine multilayers in the presence and absence of cholesterol (Mendelsohn and Taraschi, 1978). In the latter publication, Mendelsohn and Taraschi made the assumption that an increase in the

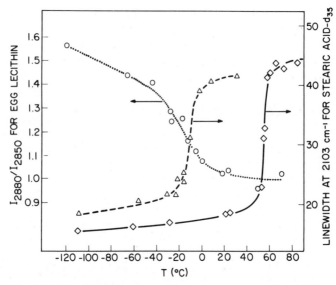

Fig. 8.15. Temperature dependence of: the linewidth of the 2103-cm^{-1} band for pure stearic acid-d_{35}, ($\Diamond$); the linewidth of the 2103-cm^{-1} band for stearic acid-d_{35} bound in lecithin multilayers ($\triangle$). The curves have their ordinate scale on the right. The linewidth measurement refers to the full width at half maximum. The intensity ratio of the C–H stretching modes at 2880 and 2850 cm^{-1} for egg lecithin multilayers ($\bigcirc$) as a function of temperature is shown with its ordinate scale on the left. [From Mendelsohn *et al.* (1976b).]

C–D linewidth is a linear function of the number of gauche rotamers formed and thus calculated the population of gauche isomers. Phase diagrams can also be constructed for these systems (Mendelsohn and Koch, 1980), and analysis of the diagrams indicates that the effect of perdeuteration on the phase behavior is not extensive.

Selective deuteration provides information regarding band assignment (Bunow and Levin, 1977b; Gaber *et al.*, 1978b; Verma and Wallach, 1977) and acts as a probe of conformation within a single phospholipid. Bansil *et al.* (1980) studied the temperature dependence of the Raman spectra of 1,2-dimyristoyl-*sn*-glycero-3-phosphocholines specifically deuterated in chain 2 at positions 3, 4, 6, 10, 12, and 14. They showed that the frequencies of the CD$_2$ stretching modes depend on the position of the CD$_2$ label, being maximum at position 3 and decreasing until they become constant beyond position 6. The degree of the observed increase in the width of the C–D bands at T$_m$ is also position dependent. An interesting example of selective deuteration is the research by Gaber *et al.* (1978c) which involved deuterating chain 1 or chain 2 of dipalymitoyl phosphatidylcholine. Differences in the spectral characteristic of the compounds containing the perdeuterated chain 1 or the perdeuterated chain 2 at a certain temperature were noted. These differences were attributed to non-equivalent conformations of the fatty acid chains at positions 1 and 2; below the pretransition, chain 2 appears to depart slightly more from the all-trans structure than does chain 1.

5. Fatty-acid chains containing an olefinic linkage

Although the biological implications of an olefin group in fatty-acid hydrocarbon chains are not fully understood, a variety of physical techniques have established that these chains have physical properties markedly different from those of the saturated chains. For example, compared to the saturated hydrocarbon chains those with an olefinic linkage have increased fluidity and pronounced differences in order and mobility. However, these effects have been little studied by Raman spectroscopy. Early vibrational studies established that cis–trans isomers about the double bond can be distinguished on the basis of their —C≡C— stretching frequencies. For cis isomers $\nu_{C=C}$ occurs near 1655 cm^{-1}, while for trans isomers $\nu_{C=C}$ is found near 1670 cm^{-1} (Davies *et al.*, 1972; Bailey and Horvat, 1972). In an early publication dealing with unsaturated fatty acids and phospholipids, Lippert and Peticolas (1972) assigned two intense features in the C–C skeletal stretching region to modes belonging to the methylene chains on either side of the double bond. Koyama and Ikeda (1980) published a detailed analysis of the Raman spectra of cis-unsaturated fatty-acid chains. The compounds studied included oleic and petroselenic acids and dioleoyl- and dipetroselinoyl-L-α-phosphatidylcholines. Koyama and Ikeda (1980) ascribed Raman spectra differences between different crystalline forms as being caused by rotational isomers around the C–C bonds immediately adjacent to the double bond. The skeletal stretching vibrations of the polymethylene chains on either side of the —C≡C— linkage appeared to be localized within each chain. This provided part of the rational for the earlier finding of Lippert and Peticolas that these polymethylene chains make separate contributions to the C–C stretching regions. Moreover, Koyama and Ikeda (1980) point out that the intensities of the 1650-cm^{-1} band and the 1270-cm^{-1} band (attributable to C–H in-plane deformations) are good probes for conformation about the $C(sp^2)$–$C(sp^3)$ axis.

6. Effect of perturbants such as cholesterol and cations on lipid properties

The conformationally sensitive bands in the C–C skeletal stretching and C–H stretching regions can be used to follow the effects of perturbants on lipid properties. The addition of cholesterol to phospholipids is a prime example and has been widely studied (e.g., Faiman *et al.*, 1976, and references therein). For cholesterol-containing phospholipids there is a marked broadening of the phase transition indicating a loss of cooperativity. In the case of dipalmitoyl lecithin monohydrate (Lippert and Peticolas, 1971), the addition of cholesterol in a 1 : 1 molar ratio fluidizes the close-packed structure below T_m by causing trans–gauche isomerizations. However, above T_m the results indicated increased rigidification of the lipid structure. In another study of dipalmitoyl lecithin multilayers, Szalontai (1976) was able to show that added benzene induced a gel-to-liquid–crystalline-phase transition well below the transition temperature usually observed.

Lis *et al.* (1975) examined the effect of several ions on phospholipid structure. By following intensity patterns in the 1100-cm^{-1} region it was inferred that dipositive ions decrease the proportions of gauche rotamers in aqueous lecithin, with the relative effect being $Ba^{2+} < Mg^{2+} < Ca^{2+} = Cd^{2+}$. Singly charged cations and anions showed little effect. The same authors also examined the effect of preparing dispersions in aqueous KI/I_2 and noted an increase in randomness in the lipid chains. Loshchilova and Karvaly (1977) have also studied interactions involving iodine and iodide. They concluded that phospholipid bilayers interact with I_2 rather than I_3^- and found no evidence for the existence of triiodide chains in iodine-doped multilayers.

B. Lipid–Protein Interactions

The first reports of Raman studies of systems that were prototypes for lipid–protein interactions appeared in 1973 and 1974. Since that time the models have become increasingly sophisticated and are approaching natural lipid–protein complexes. In an early publication Larsson and Rand (1973) discussed the Raman spectra of a coprecipitate involving the protein insulin and sodium monodecylphosphate. The spectrum of the coprecipitate in the 1100-cm^{-1} region was different from the spectrum of the sum of the constituents in a way that suggested that an increase in the number of gauche isomers in the hydrocarbon chains occurs on coprecipitation. A similar conclusion was reached by Chen *et al.* (1974) in a study involving the coprecipitation of lysozyme with sodium dodecyl sulfate (SDS). The spectrum of pure solid SDS is that of a rigid all-trans structure, whereas the SDS features in the spectrum of the coprecipitate indicated that bends were formed in the hydrocarbon chains. The backbone structure of the protein was also found to change in the coprecipitate with a loss of helical regions occurring. In a recent Raman study Lippert *et al.*

(1980) also noted a change in lysozyme backbone structure in complexes involving lysozyme and phospholipid liposomes.

In other studies using prototypes for lipid–protein complexes, Lis *et al.* (1976a,b) followed the effect of various amino acids, polypeptides, and proteins on phosphatidylcholine. The 1100- and 2900-cm^{-1} regions of the Raman spectrum were used to monitor changes in the hydrocarbon chains. Lis and co-workers (1976b) provide evidence for increased membrane fluidity in the presence of several amino acids, and, in the case of dimyristoyl phosphatidylcholine, in the presence of the polypeptides valinomycin and alamethecin. However, the latter compounds produced no changes in the fluidity of dipalmitoyl phosphatidylcholine. Later, Lis *et al.* (1976a) examined the effect on lipid structure of a variety of intrinsic and extrinsic membrane proteins. Cytochrome *c* and cytochrome *c* oxidase were considered to be examples of extrinsic and intrinsic membranes proteins, respectively. Evidence that cytochrome *c* and cytochrome *c* oxidase interacted with the membrane in a different fashion was adduced from the fact that they differentially perturbed in C–H regions in the Raman spectra of dimyristoyl and dipalmitoyl phosphatidylcholines.

Several groups (Faiman and Long, 1976; Chapman *et al.*, 1977, Weidekamm *et al.*, 1977; Susi *et al.*, 1979) have used Raman spectroscopy to investigate the interaction between the polypeptide Gramicidin A and phosphatidylcholines. In general, the Raman studies show that above T_m for the lipid, Gramicidin A causes a marked decrease in the number of gauche rotamers in the polymethylene chains, and below T_m the data suggest a slight fluidization of the chains. In another study of polypeptide–lipid association, Taraschi and Mendelsohn (1979) examined complexes of the polypeptide hormone glucagon and dimyristoyl lecithin. Glucagon interacts with the lipid when the latter is in its gel state (below 23°C) to produce a soluble complex which in some respects resembles serum lipoproteins. The Raman data for the complex at 7.5°C indicated that lateral interactions between the lipid molecules were completely disrupted and that, compared to the pure lipid, additional gauche rotamers occurred in the chains of the complexed lipid.

Two recent studies have focused on "reconstituted" protein–lipid systems in which proteins isolated from natural membranes are recombined with purified lipid systems. Curatolo *et al.* (1978) examined the effect of complexing the unusually hydrophobic protein "myelin proteolipid apoprotein" with dimyristoyl lecithin and egg lecithin. For the dimyristoyl lipid–protein complex the C–H stretching region indicated that the lipid hydrocarbon chains retain some "solid" character at temperatures of up to 18°C above the lipids T_m. In the egg-lecithin–protein complex, the lipid-phase transition becomes extremely broad covering the range $-26°$

to +15°C, with an additional small transition at 12°C being superimposed on the broad change. Curatolo *et al.* suggested that some of the egg-lecithin phospholipid is sequestered by the protein into regions that give rise to the 12°C transition. In a later study, Taraschi and Mendelsohn (1980) formed lipid–protein complexes from dipalmitoylphosphatidyl choline (and its perdeuterated analog) and glycophorin, a protein isolated from erythrocyte membranes. The temperature dependence of the C–D stretching region of the protein–lipid complexes at various molar ratios is shown in Fig. 8.16. At lipid protein molar ratio of 125:1, a broad melting event occurs with a midpoint about 15°C lower than that for the pure lipid. Using their earlier postulates concerning C–D linewidth and gauche popu-lation (Mendelsohn and Taraschi, 1978) the authors concluded that the same number of gauche isomers form in the phospholipid hydrocarbon chains during the melting process as in the phase transition of the pure lipid. There is special interest in the nature of interactions at the lipid–protein interface, and Taraschi and Mendelsohn (1980) point out that the conformation of the lipid at the interface is markedly changed by the in-teraction. Moreover, the Raman results enabled them to infer for the 125:1 sample that the perturbation region extends well beyond one shell of lipid molecules and must be at least four to five layers deep.

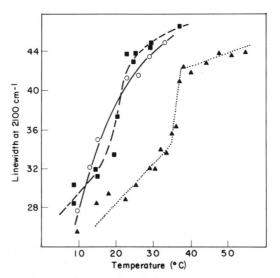

Fig. 8.16. Temperature-induced variation in the linewidth of the C–D stretching vibra-tions near 2100 cm^{-1} for D_{62} dipalmitoylphosphatidylcholine in multilayer dispersion (▲) and complexed 275:1 (■) and 125:1 (○) with glycophorin. The gel-to-liquid-crystal-phase transition appears as a sharp discontinuity near 35°C for the pure lipid. [From Taraschi and Mendelsohn (1980).]

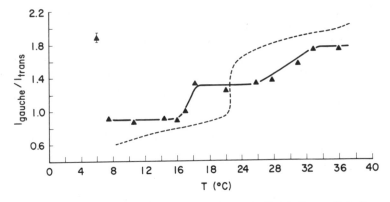

Fig. 8.17. Temperature profiles for dimyristoyl phosphatidylcholine–melittin liposomes using the I_{1090}/I_{1130} (I_{gauche}/I_{trans}) peak height intensity ratio as a marker. (—▲—) 14 : 1 lipid–melittin ratio: (---) pure lipid bilayers. [From Lavaialle *et al.* (1980).]

Lavialle *et al.* (1980) have also obtained information on the lipid–protein boundary. They studied the interaction of melittin, a polypeptide consisting of 26 amino acid residues, with dimyristoyl phosphatidyl-choline. The results illustrated in Fig. 8.17 show that for a lipid–melittin molar ratio of 14 : 1 two order–disorder transitions are observed, one above (at 29°C) and one below (at 17°C) the transition for the pure lipid (at 22.5°C). The low-temperature transition is associated with a depression of the main lipid-phase transition, and the 29°C transition is associated with the melting behavior of approximately seven immobilized boundary lipids which surround the hydrophobic portion of the melittin. In a separate study, Verma and Wallach (1976c), using different protein–lipid ratios, observed only the higher transition in their temperature profile.

C. Natural Membranes and Resonance Raman Probes of Natural Membranes

The application of Raman spectroscopy to naturally occurring mem-branes is hindered by several problems. Each of these may be overcome to a greater or lesser extent, but the difficulties account for the fact that there has been less activity in the area of natural membranes than for model systems. One difficulty is the presence of luminescence which may partially or totally obscure the Raman spectrum. Another problem is that membranes are often highly heterogeneous assemblies, and this poses a barrier to spectral assignment. A further parameter, which is not properly understood, has recently been pointed out by Forrest (1978). He demonstrated that the Raman spectra of fat-globule membranes are markedly dependent on the thermal history of the material. The rates of cooling or heating of the sample had a profound effect on the ordering of the hydrocarbon chains. This apparent "memory" or hysteresis clearly requires characterization before the type of study on phase transitions used in the model studies can be used for natural membranes.

In spite of the problems outlined, several groups have been able to obtain Raman spectra from natural membranes. The most extensively studied system is the membrane from human erythrocytes (commonly known as red blood cells) (Bulkin, 1972; Lippert *et al.*, 1975; Wallach and Verma, 1975; Verma and Wallach, 1976a,b; Goheen *et al.*, 1977). Lippert *et al.* (1975) examined the membranes from human red blood cells, which had been repeatedly washed to remove traces of fluorescent material. The Raman signatures of both protein and lipid components were observed, and, from the amide I′ (in D_2O) and amide III regions of the protein spectrum, the protein fraction was estimated to be 40–55% α helical. From an analysis of the 1100–1150-cm^{-1} region Lippert *et al.* showed that about 60% of the lipid hydrocarbon chains were in the all-trans form. Verma and Wallach have published several contributions regarding the erythrocyte system. By monitoring the C–H stretching region they observed the main gel-to-liquid-crystal transition of the lipids near −8°C (Verma and Wallach, 1976a), and they also reported evidence for a pH-sensitive transition in the physiological temperature range (Verma and Wallach, 1976b).

Schmidt-Ullrich *et al.* (1976) have compared the Raman spectra of isolated plasma membranes from resting rabbit thymocytes and cells stimulated with concanavalin A. They report changes in both lipid and protein Raman features for the stimulated cells. For sarcoplasmic reticulum membranes Milanovich *et al.* (1976b) suggest from the 1000–1150-cm^{-1} region of the spectrum that the membrane may be more fluid than those of erythrocytes. Considerable lipid fluidity was also deduced from this region in the high-quality Raman spectra of photoreceptor membranes reported by Rothschild *et al.* (1976) and of hemoglobin-free rabbit erythrocytes reported by Milanovich *et al.* (1976a).

A further approach to studying the properties and function of natural membranes is the use of RR probes, either intrinsic or extrinsic. Membrane preparations sometimes contain a trace of carotenoid material, and it has been suggested that this be used as a RR-reporter group of membrane conformation (Verma and Wallach, 1975). However, a question remains as to utility of this approach because the carotenoid may be a contaminant (Milanovich *et al.*, 1976b; Goheen *et al.*, 1977). One system in which carotenoids are known to be an important natural constituent is the membranes of photosynthetic bacteria. Koyama *et al.* (1979a) have shown that the RR spectra of carotenoids in these membranes can be used as sensitive probes of membrane potential. Changes in electrical potential cause small shifts in carotenoid absorption maxima, but, with judicious choice of RR excitation wavelength, small changes in λ_{max} can give rise to large changes in the intensity ratio of the carotenoid's ν_1 and ν_2 modes (Chapter 5, Section V). Thus the intensity ratio of ν_1 and ν_2 becomes a sensitive measure of transmembrane potential, and Koyama *et al.* (1979a)

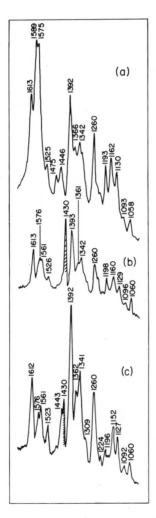

Fig. 8.18. The RR spectra of quinaldine red (a) free in solution (b) bound to *S. faecalis* in the cells resting state, and (c) after energizing cells by adding glucose to cells under (b), 441.6-nm excitation, hatched line is due to laser plasma. (Reproduced from Koyama *et al.*, 1979b, by permission.)

were able to observe oscillation in the potential across the membranes of photosynthetic cells under growing conditions. Another interesting example of the use of a polyene RR spectrum involves the extrinsic probe amphotericin B. The latter molecule is a polyene antibiotic containing seven conjugated double bonds. It was possible to follow changes in both the RR spectrum of the antibiotic and the normal Raman spectrum of lipid as a function of temperature in various lipid and lipid–cholesterol systems (Bunow and Levin, 1977c).

Amphotericin B could be regarded as a RR label, and there is another example of where a RR label has been used to follow membrane properties. The second example involves the dye quinaldine red and the intact cells of the bacterium *Streptococcus faecalis* (Koyama *et al.*, 1979b). As can be seen in Fig. 8.18, the RR spectrum of the free dye changes on binding to the membrane of the resting (nonmetabolizing) cells and changes again when the cells are provided with glucose as an energy source and the membranes become energized. In the first instance, RR-band frequency and intensity changes occur, and these effects are ascribed to ionic interactions between the dye and components of the resting membrane. Only the intensities of the RR bands change on energization, and these changes are ascribed to dye aggregation.

III. Carbohydrates

Raman spectroscopy has not been widely used for the elucidation of carbohydrate structures. A major stumbling block lies in the area of spectral interpretation; the predominance of C–C and C–O linkages in carbohydrate molecules and the similar mechanical properties of these bonds give rise to a situation where the normal modes are delocalized with contributions from many nuclei. In other words, the group frequency approach to interpretation is of little value; instead, detailed vibrational calculations have to be undertaken to establish a correlation between carbohydrate spectra and structure.

Early vibrational studies of carbohydrates were almost entirely conducted by infrared spectroscopy, and this work has been reviewed by Parker (1971). Apart from some reports on lactose (Susi and Ard, 1974) and hyaluronic acid and its constituents (She *et al.*, 1974; Tu *et al.*, 1977; Barrett and Peticolas, 1979), the majority of the recent Raman studies have been by Atalla and co-workers at the Institute of Paper Chemistry in Wisconsin, and by Blackwell and co-workers at Case Western Reserve University in Ohio. Both groups have undertaken a series of studies beginning with small "model" carbohydrates and have used information on these to interpret the Raman spectra of cellulose. Vasko *et al.* (1971)

began by identifying the O–H and C–H related vibrational modes for D-glucose, maltose, cellobiose, and dextran by various deuterium substitutions and used this data to undertake a normal coordinate analysis of α-D-glucose (Vasko *et al.*, 1972). Further vibrational calculations of β-D-glucose were performed by Cael *et al.* (1974), who were able to account for the 840-cm^{-1} and 898-cm^{-1} bands, characteristic of α-D-glucose and β-D-glucose, respectively, in terms of sensitivity of the normal modes to anomeric configuration. Cael *et al.* (1973) also studied the Raman spectra of polymorphic forms of amylose. Small band shifts were noted near 1260 and 946 cm^{-1} on conversion of B amylose to the V form. Cael *et al.* were able to support these observations and the proposed mechanism for the B–V transition by a normal coordinate analysis (Cael *et al.*, 1975b).

Atalla and co-workers began their series of papers on the vibrational spectra of saccharides by an infrared, Raman, and normal coordinate study of 1,5-anhydropentitols (Pitzner and Atalla, 1975). Since that time most of the Wisconsin group's published work has centered on cellulose, although detailed studies on the vibrational spectra of, for example, inositols (Williams, 1977) and glucose, galactose, and mannose (Wells, 1977) exist in thesis form.

As mentioned earlier, both the Case Western and Wisconsin groups have addressed the problem of cellulose structure. Cellulose is a linear homopolysaccharide of 1,4-linked β-D-glucose residues. X-ray and electron-diffraction studies have shown that native cellulose (form I) possesses an extended chain conformation with two glucose residues repeating every 10.38 Å. Cellulose I can be transformed to other polymorphic forms by swelling or precipitation from solution (to give form II), by treatment with ammonia (to give form III), and by treatment with glycerol (to give form IV). The Raman spectra, taken with other evidence, have been used to probe the differences in the conformations and packing in the different forms of cellulose. Figure 8.19 compares the Raman spectra of highly crystalline celluloses I and II. Marked differences appear in the spectra of the two forms especially in the region below 700 cm^{-1}. Atalla (Atalla and Dimick, 1975; Atalla and Nagel, 1974; Atalla, 1976; Atalla *et al.*, 1977) prefers to ascribe these differences to conformations of the individual 1,4-β-D-glucose chains. He supports his interpretation with the postulate, based on potential energy calculations, that there are just two stable conformers for cellulose chains. This postulate then leads to the prediction that the Raman spectra of celluloses III and IV can be reproduced by a weighted addition of the spectra due to forms I and II, and there is experimental evidence for this being the case (Atalla, 1976; Atalla *et al.*, 1977). Moreover, Atalla (1979) suggests that the conformational differences between forms I and II involve rotations about the glycosidic linkage. However, another interpretation for the spectral differences be-

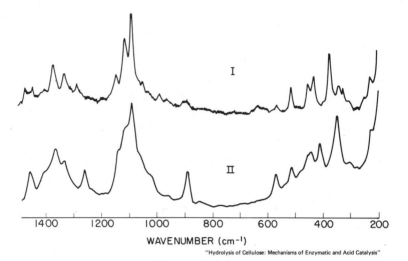

"Hydrolysis of Cellulose: Mechanisms of Enzymatic and Acid Catalysis"

Fig. 8.19. Raman spectra of highly crystalline samples of celluloses I and II. [From Atalla (1979, p. 59).]

tween celluloses I and II is preferred by Blackwell and Koenig (Blackwell, 1977; Cael *et al.*, 1975a). They emphasize the similarities of the conformations of celluloses I and II derived from X-ray diffraction methods and point out that the major differences in the Raman spectra exist in the O–H stretching region and the region below 800 cm^{-1}. They assign these differences to variations in interchain hydrogen bonding and packing rather than major conformational changes within each chain. However, Atalla (1979) states that departures from a twofold helix conformation need not be very large to account for the differences in spectra seen in Fig. 8.19. It is possible that the interpretations of both groups may be unified in a model that contains, for celluloses I and II, differences in chain packing and interchain hydrogen bonding and modest conformational chains within each chain.

Suggested Reviews for Further Reading

Krimm, S. (1978). Raman studies of the longitudinal acoustic mode in polymers. *Indian J. Pure Appl. Phys.* **16**, 335.

Lord, R. C., and Mendelsohn, R. (1980). Raman spectroscopy of membrane constituents and related molecules. *Mol. Biol., Biochem. Biophys.* **31**, 377.

Vergoten, G., Fleury, G., and Moschetto, Y. (1978). Low frequency vibrations of molecules with biological interest. *Adv. Infrared Raman Spectrosc.* **4**, 195.

Wallach, D. F. H., Verma, S. P., and Fookson, J. (1979). Application of laser Raman and infrared spectroscopy to the analysis of membrane structure. *Biochim. Biophys. Acta* **559**, 153.

Suggestions for Further Reading

As stated in the Preface, the aims of this book are twofold. One purpose is to give the biochemist a grounding in the basics of Raman spectroscopy and to demonstrate the potential of the Raman and RR techniques for biochemical research. The second purpose is to provide the spectroscopist or physical chemist with a grasp of macromolecular systems and to provide a survey which can act as a springboard for new ideas and research. These aims can be met in a single volume only by presenting the spectroscopy and biochemistry at a fairly basic level. The interested student and research worker will eventually need more in-depth coverage of both topics. This can be achieved by delving into books, review articles, and eventually research journals. Specific review articles dealing with rather narrow topics, e.g., Raman spectroscopy of lipids and membranes, RR spectroscopy of heme proteins, and the RR spectroscopy of visual pigments, have been listed at the end of the relevant chapters. Here we shall list continuing series in which future reviews are likely to be found.

Books

There are several excellent texts which provide comprehensive coverage of biochemistry, e.g.,

Lehninger, A. L. (1975). "Biochemistry." Worth, New York.
Stryer, L. (1981). "Biochemistry." Freeman, San Francisco, California.

Metzler, D. E. (1977). "Biochemistry: The Chemical Reactions of Living Cells." Academic Press, New York.

Some of the more useful books dealing with the spectroscopic aspects are

Colthup, N. B., Daly, L. H., and Wiberley, S. E. (1975). "Introduction to Infrared and Raman Spectroscopy," 2nd ed. Academic Press, New York. A good source of information on group frequencies.

Bellamy, L. J. (1975). "The Infrared Spectra of Complex Molecules," 3rd ed. Chapman & Hall, London; Bellamy, L. J. (1968). "Advances in Infrared Group Frequencies." Methuen, London. These two volumes are standard treatises for deriving chemical information from vibrational spectra.

Pinchas, S., and Lauchlit, L. (1971). "Infrared Spectra of Labelled Compounds." Academic Press, New York. A valuable source on the infrared spectra of isotopically substituted molecules prior to 1970.

Woodward, L. A. (1972). "Introduction to the Theory of Molecular Vibrations." Oxford Univ. Press, London and New York.

Parker, F. S. (1971). "Applications of Infrared Spectroscopy in Biochemistry, Biology and Medicine." Plenum, New York. Extensive coverage of the pre-1970 literature concerning biochemically important molecules.

Jaffé, H. H., and Orchin, M. (1962). "Theory and Applications of Ultraviolet Spectroscopy." Wiley, New York. This book deals with many of the chromophores encountered in RR studies and thus serves as a valuable aid for the RR work.

Szymanski, H. A., ed. (1967, 1970). "Raman Spectroscopy Theory and Practice," Vols. 1 and 2. Plenum, New York. These contain some important early reviews.

Review Series

Some of the future reviews germane to the study of biochemical systems by Raman spectroscopy are likely to be found in the following continuing series:

Advances in Infrared and Raman Spectroscopy (R. J. H. Clark and R. E. Hester, eds.),
Vibrational Spectra and Structure (J. R. Durig, ed.),
Annual Review of Biochemistry,
Annual Review of Biophysics and Bioengineering,
Quarterly Reviews of Biophysics,
CRC Critical Reviews in Biochemistry,
Chemical and Biochemical Applications of Lasers.

Journals

Unless you are fortunate enough to have access to a major library and a computerized literature search system, it is virtually impossible to keep abreast of every source of a potentially useful publication. However, by scanning the major broad-interest journals and identifying the few specialist journals in the research area of special relevance, most of the literature can be covered most of the time. For example, anybody studying the vibrational properties of lipids would probably consult some of the general interest journals listed below and, in addition, keep abreast of publications such as *The Chemistry and Physics of Lipids*. The following journals, many of which cater to a rather general readership, currently publish the majority of work appearing in the area of Raman spectroscopy and biochemistry:

Biochemically and Chemically Oriented Journals

Journal of the American Chemical Society,
Biochemistry,
Biochimica et Biophysica Acta,
Journal of Biological Chemistry,
Biopolymers,
Journal of Molecular Biology,
Proceedings of the National Academy of Sciences of the U.S.A.,
FEBS Letters,
Biochemical and Biophysical Research Communications,
Biophysical Journal.

In addition, Raman-based papers occasionally appear in *Science* (*Washington, D.C.*) and *Nature* (*London*), and these journals provide an excellent means of keeping abreast with scientific developments in general.

Spectroscopically Oriented Journals.

Spectrochimica Acta B,
Journal of Molecular Structure,
Journal of Raman Spectroscopy,
Journal of Chemical Physics,
Journal of Physical Chemistry.

References

Abe, M., Kitagawa, T., and Koygoku, Y. (1978). *J. Chem. Phys.* **69**, 4526.

Adar, F., and Erecińska, M. (1979). *Biochemistry* **18**, 1825.

Adman, E. T., Stenkamp, R. E., Sieker, L. C., and Jensen, L. H. (1978). *J. Mol. Biol.* **123**, 35.

Agalidis, I., Lutz, M., and Reiss-Husson, F. (1980). *Biochim. Biophys. Acta* **589**, 264.

Akutsu, H., and Kyogoku, Y. (1975). *Chem. Phys. Lipids* **14**, 113.

Albrecht, A. C., and Hutley, M. C. (1971). *J. Chem. Phys.* **55**, 4438.

Ambrose, E. J., and Elliott, A. (1951). *Proc. R. Soc. London, Ser. A* **208**, 75.

Andreeva, N. S., Bordsov, V. V., Goborum, N. N., Melikada, V. R., Raitz, V. S., Rostovtski, V. A., and Suutskeva, N. E. (1970). *Dokl. Akad. Nauk SSSR* **192**, 216.

Asher, I. M., Carew, E. B., and Stanley H. E. (1976). *In* "Physiology of Smooth Muscle" (E. Bulbring and M. F. Shuba, eds.), pp. 229–238. Raven, New York.

Asher, I. M., Phillies, G. D., Kim, B. J., and Stanley, H. E. (1977a). *Biopolymers* **16**, 157.

Asher, I. M., Rothschild, K. J., Anastassakis, E., and Stanley, H. E. (1977b). *J. Am. Chem. Soc.* **99**, 2024.

Asher, S. A., and Schuster, T. M. (1979). *Biochemistry* **18**, 5377.

Asher, S. A., Vickery, L. E., Schuster, T. M., and Sauer, K. (1977c). *Biochemistry* **16**, 5849.

Ashikawa, I., and Itoh, K. (1979). *Biopolymers* **18**, 1859.

Ataka, M., and Tanaka, S. (1979). *Biopolymers* **18**, 507.

Atalla, R. H. (1976). *Appl. Polym. Symp.* **28**, 659.

Atalla, R. H. (1979). *In* "Hydrolysis of Cellulose: Mechanisms of Enzymatic and Acid Catalysis" (R. Brown and L. Jurasek, eds.), p. 55. Advances in Chemistry Series No. 181. Am. Chem. Soc., Washington, D.C.

Atalla, R. H., and Dimick, B. E. (1975). *Carbohydr. Res.* **39**, C1.

Atalla, R. H., and Nagel, S. C. (1974). *Science (Washington, D.C.)* **185**, 522.

Atalla, R. H., Dimick, B. E., and Nagel, S. E. (1977). *In* "Cellulose Chemistry and Technology" (J. C. Arthur, Jr., ed.), p. 30. Am. Chem. Soc., Washington, D.C.

238

Aton, B., Doukas, A. G., Callender, R. H., Becher, B., and Ebrey, T. G. (1977). *Biochemistry* **16**, 2995.

Aton, B., Callender, R. H., and Honig, B. (1978). *Nature (London)* **273**, 784.

Aune, A. (1968). Ph.D. Thesis, Duke Univ., Durham, North Carolina.

Aylward, N. N., and Koenig, J. L. (1970). *Macromolecules* **3**, 590.

Babcock, G. T., and Salmeen, I. (1979). *Biochemistry* **18**, 2493.

Bailey, G. F., and Horvat, R. J. (1972). *J. Am. Oil Chem. Soc.* **49**, 494.

Bailey, G. S., Lee, J., and Tu, A. T. (1979). *J. Biol. Chem.* **254**, 8922.

Bandekar, J., and Krimm, S. (1979). *Proc. Natl. Acad. Sci. U.S.A.* **76**, 774.

Bandekar, J., and Krimm, S. (1980). *Biopolymers* **19**, 31.

Bansil, R., Day, J., Meadows, M., Rice, D., and Oldfield, E. (1980). *Biochemistry* **19**, 1938.

Baret, J. F., Carbone, G. P., and Sturm, J. (1979). *J. Raman Spectrosc.* **8**, 291.

Barlow, C. H., Maxwell, J. C., Wallace, W. J., and Caughey, W. S. (1973). *Biochem. Biophys. Res. Commun.* **55**, 91.

Barrett, T. W., and Peticolas, W. L. (1979). *J. Raman Spectrosc.* **8**, 35.

Barron, L. D. (1975). *Nature (London)* **257**, 372.

Barron, L. D. (1977). *Chem. Phys. Lett.* **46**, 579.

Barron, L. D. (1978). *Adv. Infrared Raman Spectrosc.* **4**, 271.

Bastian, E. J., and Martin, R. B. (1973). *J. Phys. Chem.* **77**, 1129.

Behringer, J. (1967). *In* "Raman Spectroscopy Theory and Practice" (H. A. Szymanski, ed.), Chap. 6. Plenum, New York.

Bellocq, A. M., Lord, R. C., and Mendelsohn, R. (1972). *Biochim. Biophys. Acta* **257**, 280.

Benecky, M., Li, T. Y., Schmidt, J., Frerman, F., Watters, K. L., and McFarland, J. (1979). *Biochemistry* **18**, 3471.

Bieker, L., and Schmidt, H. (1979). *FEBS Lett.* **106**, 268.

Blackwell, J. (1977). *In* "Cellulose Chemistry and Technology" (J. C. Arthur, Jr., ed.), p. 206. Am. Chem. Soc., Washington, D.C.

Blazej, D. C., and Peticolas, W. L. (1977). *Proc. Natl. Acad. Sci. U.S.A.* **74**, 2639.

Blow, D. M. (1976). *Acc. Chem. Res.* **9**, 145.

Blum, H., Adar, F., Salerno, J. C., and Leigh, J. S. (1977). *Biochem. Biophys. Res. Commun.* **77**, 650.

Blundell, T., Dodson, G., Hodgkin, D., and Mercola, D. (1972). *Adv. Protein Chem.* **26**, 279.

Bocian, D. F., Lemley, A. T., Petersen, N. O., Brudvig, G. W., and Chan, S. I. (1979). *Biochemistry* **18**, 4396.

Bosworth, Y. M., Clark, R. J. H., and Turtle, P. C. (1975). *J. Chem. Soc. Dalton Trans.* p. 2027.

Bridoux, M., and Delhaye, M. (1976). *Adv. Infrared Raman Spectrosc.* **2**, 140.

Brown, E. B., and Peticolas, W. L. (1975). *Biopolymers* **14**, 1259.

Brown, K. G., Erfurth, S. C., Small, E. W., and Peticolas, W. L. (1972a). *Proc. Natl. Acad. Sci. U.S.A.* **69**, 1467.

Brown, K. G., Kiser, E. J., and Peticolas, W. L. (1972b). *Biopolymers* **11**, 1855.

Brown, K. G., Peticolas, W. L., and Brown, E. (1973). *Biochem. Biophys. Res. Commun.* **54**, 358.

Brown, K. G., Brown, E., and Person, W. B. (1977). *J. Am. Chem. Soc.* **99**, 3128.

Brunner, H. (1974). *Naturwissenschaften* **61**, 129.

Brunner, H., and Holz, M. (1975). *Biochim. Biophys. Acta* **379**, 408.

Brunner, H., Mayer, A., and Sussner, H. (1972). *J. Mol. Biol.* **70**, 153.

Bulkin, B. J. (1972). *Biochim. Biophys. Acta* **274**, 649.

Bulkin, B. J. (1976). *Appl. Spectrosc.* **30**, 261.

Bulkin, B. J., and Krishnamachari, N. (1972). *J. Am. Chem. Soc.* **94**, 1109.

Bulkin, B. J., and Krishnan, K. (1971). *J. Am. Chem. Soc.* **93**, 5998.

Bull, C., Bullou, D. P., and Salmeen, I. (1979). *Biochem. Biophys. Res. Commun.* **87**, 836.

Bunow, M. R., and Levin, I. W. (1977a). *Biochim. Biophys. Acta* **487**, 388.

Bunow, M. R., and Levin, I. W. (1977b). *Biochim. Biophys. Acta* **489**, 191.

Bunow, M. R., and Levin, I. W. (1977c). *Biochim. Biophys. Acta* **464**, 202.

Cael, J. J., Koenig, J. L., and Blackwell, J. (1973). *Carbohydr. Res.* **29**, 123.

Cael, J. J., Koenig, J. L., and Blackwell, J. (1974). *Carbohydr. Res.* **32**, 79.

Cael, J. J., Gardner, K. H., Koenig, J. L., and Blackwell, J. (1975a). *J. Chem. Phys.* **62**, 1145.

Cael, J. J., Koenig, J. L., and Blackwell, J. (1975b). *Biopolymers* **14**, 1885.

Callender, R. H., and Honig, B. (1977). *Annu. Rev. Biophys. Bioeng.* **6**, 33.

Callender, R. H., Doukas, A., Crouch, R., and Nakanishi, K. (1976). *Biochemistry* **15**, 1621.

Cameron, D. G., Casal, H. L., and Mantsch, H. H. (1980). *Biochemistry* **19**, 3665.

Carew, E. B., Asher, I. M., and Stanley, H. E. (1975). *Science (Washington, D.C.)* **188**, 933.

Carey, P. R. (1978). *Q. Rev. Biophys.* **11**, 309.

Carey, P. R., and King, R. W. (1979). *Biochemistry* **18**, 2834.

Carey, P. R., and Salares, V. R. (1980). *Adv. Infrared Raman Spectrosc.* **7**, 1.

Carey, P. R., and Schneider, H. (1974). *Biochem. Biophys. Res. Commun.* **57**, 831.

Carey, P. R., and Schneider, H. (1976). *J. Mol. Biol.* **102**, 679.

Carey, P. R., and Schneider, H. (1978). *Acc. Chem. Res.* **11**, 122.

Carey, P. R., and Young, N. M. (1974). *Can. J. Biochem.* **52**, 273.

Carey, P. R., Schneider, H., and Bernstein, H. J. (1972). *Biochem. Biophys. Res. Commun.* **47**, 588.

Carey, P. R., Froese, A., and Schneider, H. (1973). *Biochemistry* **12**, 2198.

Carey, P. R., Carriere, R. G., Lynn, K. R., and Schneider, H. (1976). *Biochemistry* **15**, 2387.

Carey, P. R., Carriere, R. G., Phelps, D. J., and Schneider, H. (1978). *Biochemistry* **17**, 1081.

Carreira, L. A., Maguire, T. C., and Malloy, T. B. (1977). *J. Chem. Phys.* **66**, 2621.

Casal, H. L., Cameron, D. G., Smith, I. C. P., and Mantsch, H. H. (1980). *Biochemistry* **19**, 444.

Caspar, D. L. D., Cohen, C., and Longley, W. (1969). *J. Mol. Biol.* **41**, 87.

Champion, P. M., Gunsalus, I. C., and Wagner, G. C. (1978). *J. Am. Chem. Soc.* **100**, 3743.

Chapman, D., Cornell, B. A., Eliasz, A. W., and Perry, A. (1977). *J. Mol. Biol.* **113**, 517.

Chen, J. T., Shen, S. T., Chung, C. S., Chang, H., Wang, S. M., and Li, N. C. (1979). *Biochemistry* **18**, 3097.

Chen, M. C., and Lord, R. C. (1974). *J. Am. Chem. Soc.* **96**, 4750.

Chen, M. C., and Lord, R. C. (1976a). *Biochemistry* **15**, 1889.

Chen, M. C., and Lord, R. C. (1976b). *J. Am. Chem. Soc.* **98**, 990.

Chen, M. C., and Thomas, G. J. (1974). *Biopolymers* **13**, 615.

Chen, M. C., Lord, R. C., and Mendelsohn, R. (1974). *J. Am. Chem. Soc.* **96**, 3038.

Chen, M. C., Giegé, R., Lord, R. C., and Rich, A. (1975). *Biochemistry,* **14**, 4385.

Chen, M. C., Giegé, R., Lord, R. C., and Rich, A. (1978). *Biochemistry* **17**, 3134.

Chi, K. H. (1970). Ph.D. Thesis, Univ. Microfilms, Ann Arbor, Michigan, Order No. 70-19, 772.

Chinsky, L., and Turpin, P. Y. (1980). *Biopolymers* **19**, 1507.

Chinsky, L., Turpin, P. Y., Duquesne, M., and Brahms, J. (1975). *Biochem. Biophys. Res. Commun.* **65**, 1440.

Chinsky, L., Turpin, P. Y., Dusquesne, M., and Brahms, J. (1977). *Biochem. Biophys. Res. Commun.* **75**, 766.

Chinsky, L., Turpin, P. Y., Dusquesne, M., and Brahms, J. (1978). *Biopolymers* **17**, 1347.

Chottard, G., and Bolard, J. (1976). *Inorg. Chim. Acta* **20**, L17.

Chottard, G., and Mansuy, D. (1977). *Biochem. Biophys. Res. Commun.* **77**, 1333.

Chrisman, R. W., Mansy, S., Peresie, H. J., Ranade, A., Berg, T. A., and Tobias, R. S. (1977). *Bioinorg. Chem.* **7**, 245.

Clark, R. J. H., and Stewart, B. (1979). *Struct. Bonding (Berlin)* **36**, 1.

Collins, D. W., Champion, P. M., and Fitchen, D. B. (1976). *Chem. Phys. Lett.* **40**, 416.

Colman, P. M., Freeman, H. C., Guss, J. M., Murata, M., Norris, V. A., Ramshaw, J. A. M., and Venkatappa, M. P. (1978). *Nature (London)* **272**, 319.

Colthup, N. B., Daly, L. H., and Wiberley, S. E. (1975). "Introduction to Infrared and Raman Spectroscopy," 2nd ed. Academic Press, New York.

Coppey, M., Tourbez, H., Valat, P., and Alpert, B. (1980). *Nature (London)* **284**, 568.

Cotton, F. A. (1971). "Chemical Applications of Group Theory." 2nd ed. Wiley (Interscience), New York.

Cotton, F. A., and Wilkinson, G. (1980). "Advanced Inorganic Chemistry: A Comprehensive Text," 4th ed. Wiley (Interscience), New York.

Cotton, T. M., and Van Duyne, R. P. (1978). *Biochem. Biophys. Res. Commun.* **82**, 424.

Craig, W. S., and Gaber, B. P. (1977). *J. Am. Chem. Soc.* **99**, 4130.

Curatolo, W., Verma, S. P., Sakura, J. D., Small, D. M., Shipley, G. G., and Wallach, D. F. H. (1978). *Biochemistry* **17**, 1802.

Dallinger, R. F., Nestor, J. R., and Spiro, T. G. (1978). *J. Am. Chem. Soc.* **100**, 6251.

Dallinger, R. F., Guanci, J. J., Woodruff, W. H., and Rodgers, M. A. J. (1979). *J. Am. Chem. Soc.* **101**, 1355.

Davies, J. E. D., Hodge, P., Barve, J. A., Gunstone, F. D., and Ismail, I. A. (1972). *J. Chem. Soc. Perkin Trans. 2* p. 1557.

Delabar, J. M. (1978). *J. Raman Spectrosc.* **7**, 261.

Delabar, J. M., and Guschlbauer, W. (1979). *Biopolymers* **18**, 2073.

Desbois, A., Lutz, M., and Banerjee, R. (1979). *Biochemistry* **18**, 1510.

Dhamelincourt, P. (1979). Thesis, Univ. de Lille, Lille, France.

Dobek, A., Patkowski, A., Labuda, D., and Augustyniak, J. (1975). *J. Raman Spectrosc.* **3**, 45.

Doukas, A. G., Aton, B., Callender, R. H., and Ebrey, T. G. (1978). *Biochemistry* **17**, 2430.

Duff, L. L., Appelman, E. H., Shriver, D. F., and Klotz, I. M. (1979). *Biochem. Biophys. Res. Commun.* **90**, 1098.

Dunn, J. B. R., Shriver, D. F., and Klotz, I. M. (1973). *Proc. Natl. Acad. Sci. U.S.A.* **70**, 2582.

Dunn, J. B. R., Shriver, D. F., and Klotz, I. M. (1975). *Biochemistry* **14**, 2689.

Dupaix, A., Bechet, J.-J., Yon, J., Merlin, J.-C., Delhaye, M., and Hill, M. (1975). *Proc. Natl. Acad. Sci. U.S.A.* **72**, 4223.

Durig, J. R., Dunlap, R. B., and Gerson, D. J. (1980). *J. Raman Spectrosc.* **9**, 266.

Dutta, P. K., Nestor, J. R., and Spiro, T. G. (1977). *Proc. Natl. Acad. Sci. U.S.A.* **74**, 4146.

Dutta, P. K., Nestor, J., and Spiro, T. G. (1978). *Biochem. Biophys. Res. Commun.* **83**, 209.

East, E. J., Chang, R. C. C., Yu, N.-T., and Kuck, J. F. R. (1978). *J. Biol. Chem.* **253**, 1436.

Edelman, G. E., Cunningham, B. A., Reeke, G. N., Becker, J. W., Waxdal, M. J., and Wang, J. L. (1972). *Proc. Natl. Acad. Sci. U.S.A.* **69**, 2580.

Edsall, J. T. (1936). *J. Chem. Phys.* **4**, 1.

Edsall, J. T., Martin, R. B., and Hollingworth, B. R. (1958). *Proc. Natl. Acad. Sci. U.S.A.* **44**, 505.

Eickman, N. C., Solomon, E. I., Larrabee, J. A., Spiro, T. G., and Lerch, K. (1978). *J. Am. Chem. Soc.* **100**, 6529.

Epp, O., Colman, P., Fehlhammer, Bode, W., Schiffer, M., Huber, R., and Palm, W. (1974). *Eur. J. Biochem.* **45**, 513.

Erfurth, S. C., and Peticolas, W. L. (1975). *Biopolymers* **14**, 247.

Erfurth, S. C., Kiser, E. J., and Peticolas, W. L. (1972). *Proc. Natl. Acad. Sci. U.S.A.* **69**, 938.

Erfurth, S. C., Bond, P. J., and Peticolas, W. L. (1975). *Biopolymers* **14**, 1245.

Euler, H., and Hellström, H. (1932). *Z. Phys. Chem., Abt. B* **15**, 342.

Eyring, G., and Mathies, R. (1979). *Proc. Natl. Acad. Sci. U.S.A.* **76**, 33.

Faiman, R., and Larsson, K. (1976). *J. Raman Spectrosc.* **4**, 387.

Faiman, R., and Long, D. A. (1975). *J. Raman Spectrosc.* **3**, 379.

Faiman, R., and Long, D. A. (1976). *J. Raman Spectrosc.* **5**, 87.

Faiman, R., Larsson, K., and Long, D. A. (1976). *J. Raman Spectrosc.* **5**, 3.

Fasman, G. D., Hoving, H., and Timasheff, S. N. (1970). *Biochemistry* **9**, 3316.

Fasman, G. D., Itoh, K., Liu, C. S., and Lord, R. C. *Biopolymers* **17**, 1729 (1978a).

Fasman, G. D., Itoh, K., Liu, C. S., and Lord, R. C. (1978b). *Biopolymers* **17**, 125.

Felton, R. H., and Yu, N.-T. (1978). *In* "The Porphyrins" (D. Dolphin, ed.), Vol. 3, Chap. 8. Academic Press, New York.

Felton, R. H., Cheung, L. D., Phillips, R. S., and May, S. W. (1978). *Biochem. Biophys. Res. Commun.* **85**, 844.

Ferris, N. S., Woodruff, W. H., Rorabacher, D. B., Jones, T. E., and Ochrymowycz, L. A. (1978). *J. Am. Chem. Soc.* **100**, 5939.

Ferris, N. S., Woodruff, W. H., Tennent, D. L., and McMillin, D. R. (1979). *Biochem. Biophys. Res. Commun.* **88**, 288.

Forrest, G. (1976). *J. Phys. Chem.* **80**, 1127.

Forrest, G. (1978). *Chem. Phys. Lipids* **21**, 237.

Fox, J. W., Lee, J., Amorese, D., and Tu, A. T. (1979). *J. Appl. Biochem.* **1**, 336.

Fraser, R. D. B., MacRae, T. P., Parry, D. A. D., and Suzuki, E. (1971). *Polymer* **12**, 35.

Freedman, T. B., Loehr, J. S., and Loehr, T. M. (1976). *J. Am. Chem. Soc.* **96**, 2809.

Freier, S. M., Duff, L. L., Van Duyne, R. P., and Klotz, I. M. (1979). *Biochemistry* **18**, 5372.

Friedman, J. M., Rousseau, D. L., and Adar, F. (1977). *Proc. Natl. Acad. Sci. U.S.A.* **74**, 2607.

Frushour, B. G., and Koenig, J. L. (1974a). *Biopolymers* **13**, 455.

Frushour, B. G., and Koenig, J. L. (1974b). *Biopolymers* **13**, 1809.

Frushour, B. G., and Koenig, J. L. (1975a). *Biopolymers* **14**, 2115.

Frushour, B. G., and Koenig, J. L. (1975b). *Biopolymers* **14**, 649.

Funfschilling, J., and Williams, D. F. (1976). *Appl. Spectrosc.* **30**, 443.

Gaber, B. P., and Peticolas, W. L. (1977). *Biochim. Biophys. Acta* **465**, 260.

Gaber, B. P., Miskowski, V., and Spiro, T. G. (1974). *J. Am. Chem. Soc.* **96**, 6868.

Gaber, B. P., Yager, P., and Peticolas, W. L. (1978a). *Biophys. J.* **21**, 161.

Gaber, B. P., Yager, P., and Peticolas, W. L. (1978b). *Biophys. J.* **22**, 191.

Gaber, B. P., Yager, P., and Peticolas, W. L. (1978c). *Biophys. J.* **24**, 677.

Gaber, B. P., Sheridan, J. P., Bazer, F. W., and Roberts, R. M. (1979). *J. Biol. Chem.* **254**, 8340.

Galluzzi, F., Garozzo, M., and Ricci, F. F. (1974). *J. Raman Spectrosc.* **2**, 351.

Garfinkel, D., and Edsall, J. T. (1958). *J. Am. Chem. Soc.* **80**, 3818.

Genzel, L., Keilmann, F., Martin, T. P., Winterling, G., Yacoby, Y., Fröhlich, H., and Makinen, M. W. (1976). *Biopolymers* **15**, 219.

George, W. O., and Mendelsohn, R. (1973). *Appl. Spectrosc.* **27**, 390.

Gill, D., Kilponen, R. G., and Rimai, L. (1970). *Nature (London)* **227**, 743.

Goheen, S. C., Gilman, T. H., Kauffman, J. W., and Garvin, J. E. (1977). *Biochem. Biophys. Res. Commun.* **79**, 805.

Goodwin, D. C., and Brahms, J. (1978). *Nucleic Acid Res.* **5**, 835.

Goodwin, D. C., Vergne, J., Brahms, J., Defer, N., and Kruh, J. (1979). *Biochemistry* **18**, 2057.

Gramlich, V., Herbeck, R., Schlenker, P., and Schmid, E. D. (1978). *J. Raman Spectrosc.* **7**, 101.

Gratzer, W. B., Bailey, E., and Beaven, G. H. (1967). *Biochem. Biophys. Res. Commun.* **28**, 914.

Gratzer, W. B., Beaven, G. H., Rattle, H. W. E., and Bradbury, E. M. (1968). *Eur. J. Biochem.* **3**, 276.

Guillot, J.-G., Pézolet, M., and Pallotta, D. (1977). *Biochim. Biophys. Acta* **491**, 423.

Hacker, H. H. (1968). Ph.D. Thesis, Univ. of Munich, Munich.

Hammett, L. P. (1970). "Physical Organic Chemistry," 2nd ed. McGraw-Hill, New York.

Harada, I., Sugawara, Y., Matsuura, H., and Shimanouchi, T. (1975). *J. Raman Spectrosc.* **4**, 91.

Harada, I., Takamatsu, T., Shimanouchi, T., Miyazawa, T., and Tamiya, N. (1976). *J. Phys. Chem.* **80**, 1153.

Harris, J. M., Chrisman, R. W., Lytle, F. E., and Tobias, R. S. (1976). *Anal. Chem.* **48**, 1937.

Hartman, K. A., Clayton, N., and Thomas, G. J. (1973). *Biochem. Biophys. Res. Commun.* **50**, 942.

Hartman, K. A., McDonald-Ordzie, P. E., Kaper, J. M., Prescott, B., and Thomas, G. J. (1978). *Biochemistry* **17**, 2118.

Herbeck, R., Yu, T.-J., and Peticolas, W. L. (1976). *Biochemistry* **15**, 2656.

Herskovits, T. T., and Brahms, J. (1976). *Biopolymers* **15**, 687.

Hester, R. E. (1967). *In* "Raman Spectroscopy" (H. A. Symanski, ed.), Chap. 4. Plenum, New York.

Heyde, M. E., Gill, D., Kilponen, R. G., and Rimai, L. (1971). *J. Am. Chem. Soc.* **93**, 6776.

Hillig, K. W., and Morris, M. D. (1976). *Biochem. Biophys. Res. Commun.* **71**, 1228.

Hirakawa, A. Y., and Tsuboi, M. (1975). *Science (Washington, D.C.)* **188**, 359.

Hirakawa, A. Y., Nishimura, Y., Matsumoto, T., Nakanishi, M., and Tsuboi, M. (1978). *J. Raman Spectrosc.* **7**, 282.

Hol, W. G. T., van Duijnen, P. T., and Berendsen, H. J. C. (1978). *Nature (London)* **273**, 443.

Hoskins, L. C., and Alexander, V. (1977). *Anal. Chem.* **49**, 695.

Howard, F. B., Frazier, J., and Miles, H. T. (1969). *Proc. Natl. Acad. Sci. U.S.A.* **64**, 451.

Hruby, V. J., Deb, K. K., Fox, J., Bjarnason, J., and Tu, A. T. (1978). *J. Biol. Chem.* **253**, 6060.

Hsu, S. L., Moore, W. H., and Krimm, S. (1976). *Biopolymers* **15**, 1513.

Huber, R., Kukla, D., Rühlmann, A., Epp, O., and Formanek, H. (1970). *Naturwissenschaften* **57**, 389.

Inagaki, F., Tasumi, M., and Miyazawa, T. (1974). *J. Mol. Spectrosc.* **50**, 286.

Inagaki, F., Tasumi, M., and Miyazawa, T. (1975). *J. Raman Spectrosc.* **3**, 335.

Ishizaki, H., McKay, R. H., Norton, T. R., Yasunobu, K. T., Lee, J., and Tu, A. T. (1979). *J. Biol. Chem.* **254**, 9651.

Jensen, N.-H., Wilbrandt, R., Pagsberg, P. B., Sillesen, A. H., and Hansen, K. B. (1980). *J. Am. Chem. Soc.* **102**, 7441.

Johansen, J. T., and Vallee, B. L. (1971). *Proc. Natl. Acad. Sci. U.S.A.* **68**, 2532.

Johansen, J. T., and Vallee, B. L. (1975). *Biochemistry* **14**, 649.

Johnson, B. B., and Peticolas, W. L. (1976). *Annu. Rev. Phys. Chem.* **27**, 465.

Kannan, K. K., Nostrand, B., Fridborg, K., Lövgren, S., Ohlsson, A., and Petef, M. (1975). *Proc. Natl. Acad. Sci. U.S.A.* **72**, 51.

Kartha, G., Bello, J., and Harker, D. (1967). *Nature (London)* **213**, 862.

Karvaly, B., and Loshchilova, E. (1977). *Biochim. Biophys. Acta* **470**, 492.
Keyes, W. E., Loehr, T. M., and Taylor, M. L. (1978). *Biochem. Biophys. Res. Commun.* **83**, 941.
Keyes, W. E., Loehr, T. M., Taylor, M. L., and Loehr, J. S. (1979). *Biochem. Biophys. Res. Commun.* **89**, 420.
Kiefer, W. (1973). *Appl. Spectrosc.* **27**, 253.
Kiefer, W. (1977). *Adv. Infrared Raman Spectrosc.* **3**, 1.
Kiefer, W., and Bernstein, H. J. (1971). *Appl. Spectrosc.* **25**, 500.
Kincaid, J., Stein, P., and Sprio, T. G. (1979a). *Proc. Natl. Acad. Sci. U.S.A.* **76**, 549.
Kincaid, J., Stein, P., and Spiro, T. G. (1979b). *Proc. Natl. Acad. Sci. U.S.A.* **76**, 4156.
Kint, S., and Tomimatsu, Y. (1979). *Biopolymers* **18**, 1073.
Kitagawa, T., and Nagai, K. (1979). *Nature (London)* **281**, 503.
Kitagawa, T., Iizuka, T., Saito, M., and Kyogoku, Y. (1975). *Chem. Lett.* p. 849.
Kitagawa, T., Ozaki, Y., and Kyogoku, Y. (1978). *Adv. Biophys.* **11**, 153.
Kitagawa, T., Nishina, Y., Kyogoku, Y., Yamano, T., Ohishi, N., Takai-Suzuki, A., and Yagi, K. (1979a). *Biochemistry* **18**, 1804.
Kitagawa, T., Nishina, Y., Shiga, K., Watari, H., Matsumura, Y., and Yamano, T. (1979b). *J. Am. Chem. Soc.* **101**, 3376.
Kitagawa, T., Azuma, T., and Hamaguchi, K. (1979c). *Biopolymers* **18**, 451.
Koenig, J. L., and Frushour, B. G. (1972a). *Biopolymers* **11**, 1871.
Koenig, J. L., and Frushour, B. G. (1972b). *Biopolymers* **11**, 2505.
Koenig, J. L., and Sutton, P. L. (1970). *Biopolymers* **9**, 1229.
Koenig, J. L., and Sutton, P. L. (1971). *Biopolymers* **10**, 89.
Koningstein, J. A., and Gächter, B. F. (1973). *J. Opt. Soc. Am.* **63**, 892.
Koyama, Y., and Ikeda, K.-I. (1980). *Chem. Phys. Lipids* **26**, 149.
Koyama, Y., Toda, S., and Kyogoku, Y. (1977). *Chem. Phys. Lipids* **19**, 74.
Koyama, Y., Long, R. A., Martin, W. G., and Carey, P. R. (1979a). *Biochim. Biophys. Acta* **548**, 153.
Koyama, Y., Carey, P. R., Long, R. A., Martin, W. G., and Schneider, H. (1979b). *J. Biol. Chem.* **254**, 10276.
Kraut, J. (1971). *In* "The Enzymes" (P. D. Boyer, ed.), 3rd ed., Vol. 3, p. 165. Academic Press, New York.
Kronman, M. J. (1968). *Biochem. Biophys. Res. Commun.* **33**, 535.
Kuck, J. F. R., East, E. J., and Yu, N.-T. (1976). *Exp. Eye Res.* **23**, 9.
Kumar, K., and Carey, P. R. (1975). *J. Chem. Phys.* **63**, 3697.
Kumar, K., and Carey, P. R. (1977). *Can. J. Chem.* **55**, 1444.
Kumar, K., King, R. W., and Carey, P. R. (1974). *FEBS Lett.* **48**, 283.
Kumar, K., King, R. W., and Carey, P. R. (1976). *Biochemistry* **15**, 2195.
Kumar, K., Phelps, D. J., Carey, P. R., and Young, N. M. (1978). *Biochem. J.* **175**, 727.
Kurtz, D. M., Shriver, D. F., and Klotz, I. M. (1976). *J. Am. Chem. Soc.* **98**, 5033.
Kyogoku, Y., Lord, R. C., and Rich, A. (1966). *Science (Washington, D.C.)* **154**, 518.
Lafleur, L., Rice, J., and Thomas, G. J. (1972). *Biopolymers* **11**, 2423.
Lane, M. J., and Thomas, G. J. (1979). *Biochemistry* **18**, 3839.
Lanir, A., and Yu, N.-T. (1979). *J. Biol. Chem.* **254**, 5882.
Lanir, A., Yu, N.-T., and Felton, R. H. (1979). *Biochemistry* **18**, 1656.
Larrabee, J. A., Spiro, T. G., Ferris, N. S., Woodruff, W. H., Maltese, W. A., and Kerr, M. S. (1977). *J. Am. Chem. Soc.* **99**, 1979.
Larsson, K. (1973). *Chem. Phys. Lipids* **10**, 165.
Larsson, K., and Hellgren, L. (1974). *Experientia* **30**, 481.
Larsson, K., and Rand, R. P. (1973). *Biochim. Biophys. Acta* **326**, 245.

Lavialle, F., Levin, I. W., and Mollay, C. (1980). *Biochim. Biophys. Acta* **600**, 62.

Levin, I. W. (1980). *Proc. Int. Conf. Raman Spectrosc., 7th,* p. 528. North-Holland Publ., Amsterdam.

Lewis, A., Spoonhower, J., Bogomolni, R. A., Lozier, R. H., and Stoeckenius, W. (1974). *Proc. Natl. Acad. Sci. U.S.A.* **71**, 4462.

Lewis, A., Marcus, M. A., Ehrenberg, B., and Crespi, H. (1978). *Proc. Natl. Acad. Sci. U.S.A.* **75**, 4642.

Lin, V. C., and Koenig, J. L. (1976). *Biopolymers* **15**, 203.

Lippert, J. L., and Peticolas, W. L. (1971). *Proc. Natl. Acad. Sci. U.S.A.* **68**, 1572.

Lippert, J. L., and Peticolas, W. L. (1972). *Biochim. Biophys. Acta* **282**, 8.

Lippert, J. L., Gorczyca, L. E., and Meiklejohn, G. (1975). *Biochim. Biophys. Acta* **382**, 51.

Lippert, J. L., Tyminski, D., and Desmeules, P. J. (1976). *J. Am. Chem. Soc.* **98**, 7075.

Lippert, J. L., Lindsay, R. M., and Schultz, R. (1980). *Biochem. Biophys. Acta* **599**, 32.

Lipscomb, W. N., Hartsuck, J. A., Reeke, G. N., Quiocho, F. A., Bethage, P. H., Ludwig, M. L., Steitz, T. A., Muirhead, H., and Coppola, J. C. (1968). *Brookhaven Symp. Biol.* No. 21, 24.

Lis, L. J., Kauffman, J. W., and Shriver, D. J. (1975). *Biochim. Biophys. Acta* **406**, 453.

Lis, L. J., Goheen, S. C., Kauffman, T. W., and Shriver, D. F. (1976a). *Biochim. Biophys. Acta* **443**, 331.

Lis, L. J., Kauffman, J. W., and Shriver, D. F. (1976b). *Biochim. Biophys. Acta* **436**, 513.

Litman, G. W., Litman, R. S., Good, R. A., and Rosenberg, A. (1973). *Biochemistry* **12**, 2004.

Loehr, J. S., Freedman, T. B., and Loehr, T. M. (1974). *Biochem. Biophys. Res. Commun.* **56**, 510.

Loehr, T. M., Keyes, W. E., and Loehr, J. S. (1981). *In* "Oxidases and Related Redox Systems" (King, T. E., Mason, H. S., and Morrison, M., eds.). Pergamon Press, New York.

Long, T. V., and Loehr, T. M. (1970). *J. Am. Chem. Soc.* **92**, 6384.

Long, T. V., Loehr, T. M., Allkins, J. R., and Lovenberg, W. (1971). *J. Am. Chem. Soc.* **93**, 1809.

Lord, R. C. (1977). *Appl. Spectrosc.* **31**, 187.

Lord, R. C., and Thomas, G. J. (1967). *Spectrochim. Acta, Part A* **23**, 2551.

Lord, R. C., and Yu, N.-T. (1970). *J. Mol. Biol.* **50**, 509.

Loshchilova, E., and Karvaly, B. (1977). *Chem. Phys. Lipids* **19**, 159.

Lowe, G., and Williams, A. (1965). *Biochem. J.* **96**, 189.

Lowey, S., Slayter, H. S., Weeds, A. G., and Baker, H. (1969). *J. Mol. Biol.* **42**, 1.

Ludwig, M. L., Hartsuck, J. A., Steitz, T. A. Muirhead, H., Coppola, J. C., Reeke, G. N., and Lipscomb, W. N. (1967). *Proc. Natl. Acad. Sci. U.S.A.* **57**, 511.

Luoma, G. A., and Marshall, A. G. (1978a). *Proc. Natl. Acad. Sci. U.S.A.* **75**, 4901.

Luoma, G. A., and Marshall, A. G. (1978b). *J. Mol. Biol.* **125**, 95.

Lutz, M. (1972). *C. R. Hebd. Seances Acad. Sci., Ser. B* **275**, 497.

Lutz, M. (1977). *Biochim. Biophys. Acta* **460**, 408.

Lutz, M. (1980). *Proc. Int. Conf. Raman Spectrosc., 7th,* p. 520. North-Holland Pub., Amsterdam.

Lutz, M., and Kleo, J. (1976). *Biochem. Biophys. Res. Commun.* **69**, 711.

Lutz, M., and Kleo, J. (1979). *Biochim. Biophys. Acta* **546**, 365.

Lutz, M., Agalidis, I., Hervo, G., Cogdell, R. J., and Reiss-Husson, F. (1978). *Biochim. Biophys. Acta* **503**, 287.

Lyons, K. B., Friedman, J. M., and Fleury, P. A. (1978). *Nature (London)* **275**, 565.

McClain, W. M. (1971). *J. Chem. Phys.* **55**, 2789.

MacClement, B. A. E., Carriere, R. G., Phelps, D. J., and Carey, P. R. (1981). *Biochemistry* **20**, 3438.

McFarland, J. T., Watters, K. L., and Petersen, R. L. (1975). *Biochemistry* **14**, 624.

Malhotra, O. P., and Bernhard, S. A. (1968). *J. Biol. Chem.* **243**, 1243.

Manfait, M., Alix, A. J. P., Jeannesson, P., Jardillier, J. C., and Theopharides, T. (1980). *Proc. Int. Conf. Raman Spectrosc., 7th,* p. 636. North-Holland Publ., Amsterdam.

Mansy, S., and Peticolas, W. L. (1976). *Biochemistry* **15**, 2650.

Mansy, S., Engstrom, S. K., and Peticolas, W. L. (1976). *Biochem. Biophys. Res. Commun.* **68**, 1242.

Marcus, M. A., and Lewis, A. (1978). *Biochemistry* **17**, 4722.

Martin, J. C., Wartell, R. M., and O'Shea, D. C. (1978). *Proc. Natl. Acad. Sci. U.S.A.* **75**, 5483.

Mathies, R. (1979). *In* "Chemical and Biochemical Applications of Lasers" (C. B. Moore, ed.), Vol. 4, p. 55. Academic Press, New York.

Mathies, R., and Yu, N.-T. (1978). *J. Raman Spectrosc.* **7**, 349.

Mathies, R., Oseroff, A. R., and Stryer, L. (1976). *Proc. Natl. Acad. Sci. U.S.A.* **73**, 1.

Mathies, R., Eyring, G., Curry, B., Bloek, A., Palings, I., Fransen, R., and Lugtenburg, J. (1980). *Proc. Int. Conf. Raman Spectrosc. 7th,* p. 546. North-Holland Publ., Amsterdam.

Maxfield, F. R., and Scheraga, H. A. (1977). *Biochemistry* **16**, 4443.

Mayer, E., Gardiner, D. J., and Hester, R. E. (1973). *Mol. Phys.* **26**, 783.

Mendelsohn, R., and Koch, C. C. (1980). *Biochim. Biophys. Acta* **598**, 260.

Mendelsohn, R., and Maisano, J. (1978). *Biochim. Biophys. Acta* **506**, 192.

Mendelsohn, R., and Taraschi, T. (1978). *Biochemistry* **17**, 3944.

Mendelsohn, R., Verma, A. L., Bernstein, H. J., and Kates, M. (1974). *Can. J. Biochem.* **52**, 774.

Mendelsohn, R., Sunder, S., and Bernstein, H. J. (1975a). *Biochim. Biophys. Acta* **413**, 329.

Mendelsohn, R., Sunder, S., Verma, A. L., and Bernstein, H. J. (1975b). *J. Chem. Phys.* **62**, 37.

Mendelsohn, R., Sunder, S., and Bernstein, H. J. (1976a). *Biochim. Biophys. Acta* **419**, 563.

Mendelsohn, R., Sunder, S., and Bernstein, H. J. (1976b). *Biochim. Biophys. Acta* **443**, 613.

Milanovich, F. P., Shore, B., Harney, R. C., and Tu, A. T. (1976a). *Chem. Phys. Lipids* **17**, 79.

Milanovich, F. P., Yeh, Y., Baskin, R. J., and Harney, R. C. (1976b). *Biochim. Biophys. Acta* **419**, 243.

Miller, F. A., and Harney, B. M. (1970). *Appl. Spectrosc.* **24**, 291.

Miskowski, V., Tang, S.-P. W., Spiro, T. G., Shapiro, E., and Moss, T. H. (1975). *J. Am. Chem. Soc.* **14**, 1244.

Miyazawa, T. (1960). *J. Chem. Phys.* **32**, 1647.

Miyazawa, T., and Blout, E. R. (1961). *J. Am. Chem. Soc.* **83**, 712.

Miyazawa, T., Shimanouchi, T., and Mizushima, S. (1958). *J. Chem. Phys.* **29**, 611.

Morikawa, K., Tsuboi, M., Takahashi, S., Kyogoku, Y., Mitsui, Y., Iitaka, Y., and Thomas, G. J. (1973). *Biopolymers* **12**, 799.

Nelson, W. H., and Carey, P. R. (1981). *J. Raman Spectrosc.* **11**, 326.

Nestor, J., and Spiro, T. G. (1973). *J. Raman Spectrosc.* **1**, 539.

Nishimura, Y., and Tsuboi, M. (1978). *Chem. Phys. Lett.* **59**, 210.

Nishimura, Y., Hirakawa, A. Y., Tsuboi, M., and Nishimura, S. (1976). *Nature (London)* **260**, 173.

Nishimura, Y., Hirakawa, A. Y., and Tsuboi, M. (1978). *Adv. Infrared Raman Spectrosc.* **5**, 217.

Nishina, Y., Kitagawa, T., Shiga, K., Horiike, K., Matsumura, Y., Watari, H., and Yamano, T. (1978). *J. Biochem.* (*Tokyo*) **84**, 925.

Nishina, Y., Kitagawa, T., Shiga, K., Watari, H., and Yamano, T. (1980). *J. Biochem.* (*Tokyo*) **87**, 831.

Nogami, N., Sugeta, H., and Miyazawa, T. (1975). *Chem. Lett.* p. 147.

O'Connor, T., Johnson, C., and Scovell, W. M. (1976a). *Biochim. Biophys. Acta* **447**, 484.

O'Connor, T., Johnson, C., and Scovell, W. M. (1976b). *Biochim. Biophys. Acta* **447**, 495.

Oesterhelt, D., and Stoeckenius, W. (1973). *Proc. Natl. Acad. Sci. U.S.A.* **70**, 2853.

Okabayashi, H., Okuyama, M., Kitagawa, T., and Miyazawa, T. (1974). *Bull. Chem. Soc. Jpn.* **47**, 1075.

Oseroff, A. R., and Callender, R. H. (1974). *Biochemistry* **13**, 4243.

Ozaki, Y., Kitagawa, T., Kyogoku, Y., Shimada, H., Iizuka, T., and Ishimura, Y. (1976). *J. Biochem.* (*Tokyo*) **80**, 1447.

Ozaki, Y., Kitagawa, T., Kyogoku, Y., Imai, Y., Hashimoto-Yutsudo, C., and Sato, R. (1978). *Biochemistry* **17**, 5826.

Ozaki, Y., King, R. W., and Carey, P. R. (1981). *Biochemistry* **20**, 3219.

Ozaki, Y., Storer, A. C., and Carey, P. R. (1982). *Can. J. Chem.* **60**, 190.

Painter, P. C., and Coleman, M. M. (1978). *Biopolymers* **17**, 2475.

Painter, P. C., and Koenig, J. L. (1975). *Biopolymers* **14**, 457.

Painter, P. C., and Koenig, J. L. (1976a). *Biopolymers* **15**, 241.

Painter, P. C., and Koenig, J. L. (1976b). *Biopolymers* **15**, 229.

Painter, P. C., and Koenig, J. L. (1976c). *Biopolymers* **15**, 2155.

Painter, P. C., and Mosher, L. E. (1979). *Biopolymers* **18**, 3121.

Parker, F. S. (1971). "Applications of Infrared Spectroscopy in Biochemistry, Biology and Medicine. Plenum, New York.

Patrick, D. M., Wilson, J. E., and Leroi, G. E. (1974). *Biochemistry* **13**, 2813.

Penner, A. P., and Siebrand, W. (1976). *Chem. Phys. Lett.* **39**, 11.

Petersen, R. L., Li, T.-Y., McFarland, J. T., and Watters, K. L. (1977). *Biochemistry* **16**, 726.

Peticolas, W. L. (1975). *Biochimie* **57**, 417.

Peticolas, W. L. (1979). *Methods Enzymol.*, **61**, 425.

Pézolet, M., Nafie, L. A., and Peticolas, W. L. (1973). *J. Raman Spectrosc.* **1**, 455.

Pézolet, M., Yu, T.-J., and Peticolas, W. L. (1975). *J. Raman Spectrosc.* **3**, 55.

Pézolet, M., Pigeon-Gosselin, M., and Coulombe, L. (1976). *Biochim. Biophys. Acta* **453**, 502.

Pézolet, M., Pigeon-Gosselin, M., and Caillé, J.-P. (1978). *Biochim. Biophys. Acta* **533**, 263.

Pézolet, M., Pigeon-Gosselin, M., Nadeu, J., and Caillé, J.-P. (1980). *Biophys. J.* **31**, 1.

Pflug, H. D., and Jaeschke-Boyer, H. (1979). *Nature* (*London*) **280**, 483.

Phelps, D. J., Schneider, H., and Carey, P. R. (1981). *Biochemistry* **20**, 3447.

Phillips, D. C. (1967). *Proc. Natl. Acad. Sci. U.S.A.* **57**, 484.

Pinchas, S., and Laulicht, I. (1971). "Infrared Spectra of Labelled Compounds." Academic Press, New York.

Pink, D. A., Green, T. J., and Chapman, D. (1980). *Biochemistry* **19**, 349.

Pitzner, L. J., and Atalla, R. H. (1975). *Spectrochim. Acta, Part A* **31**, 911.

Placzek, G. (1934). *In* "Rayleigh and Raman Scattering," UCRL Transl. No. 526 L; from "Handbuch der Radiologie" (E. Marx, ed.), Vol. VI, Part 2, pp. 209–374. Akad. Verlagsges., Leipzig.

Poljak, R. J., Amzel, L. M., Chen, B. L., Phizackerly, R. P., and Saul, F. (1974). *Proc. Natl. Acad. Sci. U.S.A.* **71**, 3440.

Prescott, B., Gamache, R., Livramento, J., and Thomas, G. J. (1974). *Biopolymers* **13**, 1821.

Que, L., and Heistand, R. H. (1979). *J. Am. Chem. Soc.* **101**, 2219.
Que, L., Heistand, R. H., Mayer, R., and Roe, A. L. (1980). *Biochemistry* **19**, 2588.
Raman, C. V., and Krishnan, K. S. (1928). *Nature (London)* **121**, 501.
Remba, R. D., Champion, P. M., Fitchen, D. B., Chiang, R., and Hager, L. P. (1979). *Biochemistry* **18**, 2280.
Rimai, L., Heyde, M. E., and Carew, E. B. (1970a). *Biochem. Biophys. Res. Commun.* **38**, 231.
Rimai, L., Kilponen, R. G., and Gill, D. (1970b). *J. Am. Chem. Soc.* **92**, 3824.
Rimai, L., Kilponen, R. G., and Gill, D. (1970c). *Biochem. Biophys. Res. Commun.* **41**, 492.
Rimai, L., Heyde, M. E., and Gill, D. (1973). *J. Am. Chem. Soc.* **95**, 4493.
Roels, H., Preaux, G., and Lontie, R. (1971). *Biochemie* **53**, 1085.
Rosasco, G. J. (1980). *Adv. Infrared Raman Spectrosc.* **7**, 223.
Rothschild, K. J., Andrew, J. R., Degrip, W. J., and Stanley, H. E. (1976). *Science (Washington, D.C.)* **191**, 1176.
Rothschild, K. J., Asher, I. M., Stanley, H. E., and Anastassakis, E. (1977). *J. Am. Chem. Soc.* **99**, 2032.
Salama, S., and Spiro, T. G. (1977). *J. Raman Spectrosc.* **6**, 57.
Salama, S., and Spiro, T. G. (1978). *J. Am. Chem. Soc.* **100**, 1105.
Salama, S., Stong, J. D., Neilands, J. B., and Spiro, T. G. (1978). *Biochemistry* **17**, 3781.
Salares, V. R., Mendelsohn, R., Carey, P. R., and Bernstein, H. J. (1976). *J. Phys. Chem.* **80**, 1137.
Salares, V. R., Young, N. M., Bernstein, H. J., and Carey, P. R. (1977a). *Biochemistry* **16**, 4751.
Salares, V. R., Young, N. M., Carey, P. R., and Bernstein, H. J. (1977b). *J. Raman Spectrosc.* **6**, 282.
Salares, V. R., Young, N. M., Bernstein, H. J., and Carey, P. R. (1979). *Biochim. Biophys. Acta* **576**, 176.
Salmeen, I., Rimai, L., and Babcock, G. (1978). *Biochemistry* **17**, 800.
Saperstein, D. D., Rein, A. J., Poe, M., and Leahy, M. F. (1978). *J. Am. Chem. Soc.* **100**, 4296.
Sasaki, K., Dockerill, S., Adamiak, D. A., Tickle, I. J., and Blundell, T. (1975). *Nature (London)* **257**, 751.
Savoie, R., Klump, H., and Peticolas, W. L. (1978). *Biopolymers* **17**, 1335.
Schachar, R. A., and Solin, S. A. (1975). *Invest. Ophthalmol.* **14**, 380.
Schachtschneider, J. H., and Snyder, R. G. (1963). *Spectrochim. Acta* **19**, 117.
Schaufele, R. F. (1968). *J. Chem. Phys.* **49**, 4168.
Schaufele, R. F., and Shimanouchi, T. (1967). *J. Chem. Phys.* **47**, 3605.
Schechter, E., and Blout, E. R. (1964). *Proc. Natl. Acad. Sci. U.S.A.* **51**, 695.
Scherer, J. R., and Snyder, R. G. (1980). *J. Chem. Phys.* **72**, 5798.
Scheule, R. K., Van Wart, H. E., Vallee, B. L., and Scheraga, H. A. (1977). *Proc. Natl. Acad. Sci. U.S.A.* **74**, 3273.
Scheule, R. K., Van Wart, H. E., Zweifel, B. O., Vallee, B. L., and Scheraga, H. A. (1979). *J. Inorg. Biochem.* **11**, 283.
Scheule, R. K., Van Wart, H. E., Vallee, B. L., and Scheraga, H. A. (1980). *Biochemistry* **19**, 759.
Schmidt-Ullrich, R., Verma, S. P., and Wallach, D. F. H. (1976). *Biochim. Biophys. Acta* **426**, 477.
Scholler, D. M., and Hoffman, B. M. (1979). *J. Am. Chem. Soc.* **101**, 1655.
She, C. Y., Dinh, N. D., and Tu, A. T. (1974). *Biochim. Biophys. Acta* **372**, 345.
Shelnutt, J. A., O'Shea, D. C., Yu, N.-T., Cheung, L. D., and Felton, R. H. (1976). *J. Chem. Phys.* **64**, 1156.

Shelnutt, J. A., Rousseau, D. L., Friedman, J. M., and Simon, S. R. (1979). *Proc. Natl. Acad. Sci. U.S.A.* **76**, 4409.

Shie, M., Dobrov, E. N., and Tikchonenko, T. I. (1978). *Biochem. Biophys. Res. Commun.* **81**, 907.

Shimanouchi, T., Tsuboi, M., and Kyogoku, Y. (1967). *Adv. Chem. Phys.* **7**, 435.

Shorygin, P. P. (1978). *Russ. Chem. Rev. (Engl. Transl.)* **47**, 907; *Usp. Khim.* **47**, 1697.

Siamwiza, M. N., Lord, R. C., Chen, M. C., Takamatsu, T., Harada, I., Matsuura, H., and Shimanouchi, T. (1975). *Biochemistry* **14**, 4870.

Siebrand, W., and Zgierski, M. Z. (1979). *In* "Excited States" (E. C. Lim, ed.), Vol. 4, Chap. 1. Academic Press, New York.

Siiman, O., and Carey, P. R. (1980). *J. Inorg. Biochem.* **12**, 353.

Siiman, O., Young, N. M., and Carey, P. R. (1974). *J. Am. Chem. Soc.* **96**, 5583.

Siiman, O., Young, N. M., and Carey, P. R. (1976). *J. Am. Chem. Soc.* **98**, 744.

Small, E. W., and Peticolas, W. L. (1971). *Biopolymers* **10**, 1377.

Small, E. W., Fanconi, B., and Peticolas, W. L. (1970). *J. Chem. Phys.* **52**, 4369.

Snyder, R. G. (1967). *J. Chem. Phys.* **47**, 1316.

Snyder, R. G., and Schachtschneider, J. H. (1963). *Spectrochim. Acta* **19**, 85.

Snyder, R. G., Hsu, S. L., and Krimm, S. (1978). *Spectrochim. Acta, Part A* **34**, 395.

Snyder, R. G., Scherer, J. R., and Gaber, B. P. (1980). *Biochim. Biophys. Acta* **601**, 47.

Sonnich Mortensen, O. (1969). *Chem. Phys. Lett.* **3**, 4.

Spaulding, L. D., Chang, C. C., Yu, N.-T., and Felton, R. H. (1975). *J. Am. Chem. Soc.* **97**, 2517.

Spiker, R. C., and Levin, I. W. (1975). *Biochim. Biophys. Acta* **388**, 361.

Spiker, R. C., and Levin, I. W. (1976a). *Biochim. Biophys. Acta* **433**, 457.

Spiker, R. C., and Levin, I. W. (1976b). *Biochim. Biophys. Acta* **455**, 560.

Spiro, T. G. (1980). *Proc. Int. Conf. Raman Spectrosc., 7th,* p. 510. North-Holland Publ., Amsterdam.

Spiro, T. G., and Burke, J. M. (1976). *J. Am. Chem. Soc.* **98**, 5482.

Spiro, T. G., and Stein, P. (1977). *Annu. Rev. Phys. Chem.* **28**, 501.

Spiro, T. G., and Strekas, T. C. (1972). *Proc. Natl. Acad. Sci. U.S.A.* **69**, 2622.

Spiro, T. G., and Strekas, T. C. (1974). *J. Am. Chem. Soc.* **96**, 338.

Spiro, T. G., Stong, J. D., and Stein, P. (1979). *J. Am. Chem. Soc.* **101**, 2648.

Stockburger, M., Klusmann, W., Gattermann, H., Massig, G., and Peters, R. (1979). *Biochemistry* **18**, 4886.

Storer, A. C., Murphy, W. F., and Carey, P. R. (1979). *J. Biol. Chem.* **254**, 3163.

Storer, A. C., Phelps, D. J., and Carey, P. R. (1981). *Biochemistry* **20**, 3454.

Storer, A. C., Ozaki, Y., and Carey, P. R. (1982). *Can. J. Chem.* **60**, 199.

Strekas, T. C., and Spiro, T. G. (1972a). *Biochim. Biophys. Acta* **263**, 830.

Strekas, T. C., and Spiro, T. G. (1972b). *Biochim. Biophys. Acta* **278**, 188.

Strekas, T. C., and Spiro, T. G. (1973). *J. Raman Spectrosc.* **1**, 387.

Strekas, T. C., Packer, A. J., and Spiro, T. G. (1973). *J. Raman Spectrosc.* **1**, 197.

Sturm, J., Savoie, R., Edelson, M., and Peticolas, W. L. (1978). *Indian J. Pure Appl. Phys.* **16**, 327.

Sufrà, S., Dellepiane, G., Masetti, G., and Zerbi, G. (1977). *J. Raman Spectrosc.* **6**, 267.

Sugawara, Y., Harada, I., Matsuura, H., and Shimanouchi, T. (1978). *Biopolymers* **17**, 1405.

Sugeta, H., Go, A., and Miyazawa, T. (1972). *Chem. Lett.* p. 83.

Sugeta, H., Go, A., and Miyazawa, T. (1973). *Bull. Chem. Soc. Jpn.* **46**, 3407.

Sulkes, M., Lewis, A., and Marcus, M. A. (1978). *Biochemistry* **17**, 4712.

Sunder, S., Mendelsohn, R., and Bernstein, H. J. (1976). *Chem. Phys. Lipids* **17**, 456.

Susi, H., and Ard, J. S. (1974). *Carbohydr. Res.* **37**, 351.

Susi, H., Sampugna, J., Hampson, J. W., and Ard, J. S. (1979). *Biochemistry* **18**, 297.

Szalontai, B. (1976). *Biochem. Biophys. Res. Commun.* **70**, 947.
Takamatsu, T., Harada, I., Shimanouchi, T., Ohta, M. and Hayashi, K. (1976). *FEBS Lett.* **72**, 291.
Takenaka, T. (1979). *Adv. Colloid Interface Sci.* **11**, 291.
Takesada, H., Nakanishi, M., Hirakawa, A. Y., and Tsuboi, M. (1976). *Biopolymers* **15**, 1929.
Tang, S.-P. W., Spiro, T. G., Mukai, K., and Kimura, T. (1973). *Biochem. Biophys. Res. Commun.* **53**, 869.
Tang, S.-P. W., Spiro, T. G., Antanaitis, C., Moss, T. H., Holm, R. H., Herskovitz, T., and Mortensen, L. E. (1975). *Biochem. Biophys. Res. Commun.* **62**, 1.
Tang, J., James, M. N. G., Hsu, I. N., Jenkins, J. A., and Blundell, T. L. (1978). *Nature (London)* **271**, 618.
Taraschi, T., and Mendelsohn, R. (1979). *J. Am. Chem. Soc.* **101**, 1050.
Taraschi, T., and Mendelsohn, R. (1980). *Proc. Natl. Acad. Sci. U.S.A.* **77**, 2362.
Tasumi, M., Shimanouchi, T., and Miyazawa, T. (1962). *J. Mol. Spectrose.* **9**, 261.
Tatsuno, Y., Saeki, Y., Iwaki, M., Yagi, T., Nozaki, M., Kitagawa, T., and Otsuka, S. (1978). *J. Am. Chem. Soc.* **100**, 4614.
Teixeira-Dias, J. C. C., Jardim-Barreto, V. M., Ozaki, Y., Storer, A. C., and Carey, P. R. (1982). *Can. J. Chem.* **60**, 174.
Terner, J., Campion, A., and El-Sayed, M. A. (1977). *Proc. Natl. Acad. Sci. U.S.A.* **74**, 5212.
Terner, J., Hsieh, C.-L., Burns, A. R., and El-Sayed, M. A. (1979). *Proc. Natl. Acad. Sci. U.S.A.* **76**, 3046.
Terner, J., Spiro, T. G., Nagumo, M., Nicol, M. F., and El-Sayed, M. A. (1980). *J. Am. Chem. Soc.* **102**, 3238.
Thamann, T. J., Loehr, J. S., and Loehr, T. M. (1977). *J. Am. Chem. Soc.* **99**, 4187.
Theimer, O. H. (1957). *J. Chem. Phys.* **27**, 408.
Thomas, G. J. (1975). *Vib. Spectra Struct.* **3**, 239.
Thomas, G. J. (1976). *Appl. Spectrosc.* **30**, 483.
Thomas, G. J., and Barylski, J. R. (1970). *Appl. Spectrosc.* **24**, 463.
Thomas, G. J., and Livramento, J. (1975). *Biochemistry* **14**, 5210.
Thomas, G. J., and Murphy, P. (1975). *Science (Washington, D.C.)* **188**, 1205.
Thomas, G. J., Medeiros, G. C., and Hartman, K. A. (1972). *Biochim. Biophys. Acta* **277**, 71.
Thomas, G. J., Chen, M. C., and Hartman, K. A. (1973a). *Biochim. Biophys. Acta* **324**, 37.
Thomas, G. J., Chen, M. C., Lord, R. C., Kotsiopoulos, P. S., Tritton, T. R., and Mohr, S. C. (1973b). *Biochem. Biophys. Res. Commun.* **54**, 570.
Thomas, G. J., Prescott, B., McDonald-Ordzie, P. E., and Hartman, K. A. (1976). *J. Mol. Biol.* **102**, 103.
Thomas, G. J., Prescott, B., and Olins, D. E. (1977). *Science (Washington, D.C.)* **197**, 385.
Thompson, J. S., Marks, T. J., and Ibers, J. A. (1979). *J. Am. Chem. Soc.* **101**, 4180.
Tiffany, M. K., and Krim, S. (1969). *Biopolymers* **8**, 347.
Tobin, M. C. (1969). *Spectrochim. Acta, Part A* **25**, 1855.
Tomimatsu, Y., and Gaffield, W. (1965). *Biopolymers* **3**, 509.
Tomimatsu, Y., Kint, S., and Scherer, J. R. (1973). *Biochem. Biophys. Res. Commun.* **54**, 1067.
Tomimatsu, Y., Kint, S., and Scherer, J. R. (1976). *Biochemistry* **15**, 4918.
Tomlinson, B. L., and Peticolas, W. L. (1970). *J. Chem. Phys.* **52**, 2154.
Tosi, L., and Garnier, A. (1979). *Biochem. Biophys. Res. Commun.* **91**, 1273.
Tosi, L., Garnier, A., Herve, M., and Steinbuch, M. (1975). *Biochem. Biophys. Res. Commun.* **65**, 100.

Townend, R., Kumosinki, T. F., and Timasheff, S. N. (1967). *J. Biol. Chem.* **242**, 4538.

Tsai, C.-W., and Morris, M. D. (1975). *Anal. Chim. Acta* **76**, 193.

Tsuboi, M., Takahashi, S., Muraishi, S., Kajiura, T., and Nishimura, S. (1971). *Science (Washington, D.C.)* **174**, 1142.

Tsuboi, M., Hirakawa, A. Y., Nishimura, Y., and Harada, I. (1974). *J. Raman Spectrosc.* **2**, 609.

Tu, A. T., Dinh, N. D., She, C. Y., and Maxwell, J. (1977). *Stud. Biophys.* **63**, 115.

Tu, A. T., Lee, J., Deb, K. K., and Hruby, V. J. (1979). *J. Biol. Chem.* **254**, 3272.

Turano, T. A., Hartman, K. A., and Thomas, G. J. (1976). *J. Phys. Chem.* **80**, 1157.

Twardowski, J. (1978). *Biopolymers* **17**, 181.

Twardowski, J. (1979). *Biochim. Biophys. Acta* **578**, 116.

Van Duyne, R. P., Jeanmaire, D. L., and Shriver, D. F. (1974). *Anal. Chem.* **46**, 213.

Van Wart, H. E., and Scheraga, H. A. (1976). *J. Phys. Chem.* **80**, 1823.

Van Wart, H. E., and Scheraga, H. A. (1978). *Methods Enzymol.* **49**, 67.

Van Wart, H. E., Lewis, A., Scheraga, H. A., and Saeva, F. D. (1973). *Proc. Natl. Acad. Sci. U.S.A.* **70**, 2619.

Vasko, P. D., Blackwell, J., and Koenig, J. L. (1971). *Carbohydr. Res.* **19**, 297.

Vasko, P. D., Blackwell, J., and Koenig, J. L. (1972). *Carbohydr. Res.* **23**, 407.

Verma, S. P., and Wallach, D. F. H. (1975). *Biochim. Biophys. Acta* **401**, 168.

Verma, S. P., and Wallach, D. F. H. (1976a). *Biochim. Biophys. Acta* **436**, 307.

Verma, S. P., and Wallach, D. F. H. (1976b). *Proc. Natl. Acad. Sci. U.S.A.* **73**, 3558.

Verma, S. P., and Wallach, D. F. H. (1976c). *Biochim. Biophys. Acta* **426**, 616.

Verma, S. P., and Wallach, D. F. H. (1977). *Biochim. Biophys. Acta* **486**, 217.

Wallach, D. F. H., and Verma, S. P. (1975). *Biochim. Biophys. Acta* **382**, 542.

Warren, C. H., and Hooper, D. L. (1973). *Can. J. Chem.* **51**, 3901.

Warshel, A. (1977). *Annu. Rev. Biophys. Bioeng.* **6**, 273.

Weidekamm, E., Bamberg, E., Brdiczka, D., Wildermuth, G., Macco, F., Lehmann, W., and Weber, R. (1977). *Biochim. Biophys. Acta* **464**, 442.

Wells, H. A. (1977). Ph.D. Thesis, Inst. Pap. Chem., Appleton, Wisconsin.

Westover, C. J., Tiffany, M. L., and Krimm, S. (1962). *J. Mol. Biol.* **4**, 316.

Williams, R. M. (1977). Ph.D. Thesis, Inst. Pap. Chem., Appleton, Wisconsin.

Williams, R. W., Dunker, A. K., and Peticolas, W. L. (1980a). *Biophys. J.* **32**, 232.

Williams, R. W., Cutrera, T., Dunker, A. K., and Peticolas, W. L. (1980b). *FEBS Lett.* **115**, 306.

Wilson, E. B. (1934). *Phys. Rev.* **45**, 706.

Woodruff, W. H., and Farquharson, S. (1978). *Science (Washington, D.C.)* **201**, 831.

Wozniak, W. T., and Spiro, T. G. (1973). *J. Am. Chem. Soc.* **95**, 3402.

Yamamoto, T., Palmer, G., Gill, D., Salmeen, I. T., and Rimai, L. (1973). *J. Biol. Chem.* **248**, 5211.

Yaney, P. P. (1976). *J. Raman Spectrosc.* **5**, 219.

Yellin, N., and Levin, I. W. (1977a). *Biochim. Biophys. Acta* **468**, 490.

Yellin, N., and Levin, I. W. (1977b). *Biochemistry* **16**, 642.

Yoshizawa, T., and Wald, G. (1963). *Nature (London)* **197**, 1279.

Yu, N.-T. (1974). *J. Am. Chem. Soc.* **96**, 4664.

Yu, N.-T. (1977). *CRC Crit. Rev. Biochem.* **4**, 229.

Yu, N.-T., and East, E. J. (1975). *J. Biol. Chem.* **250**, 2196.

Yu, N.-T., and Jo, B. H. (1973a). *Arch. Biochem. Biophys.* **156**, 469.

Yu, N.-T., and Jo, B. H. (1973b). *J. Am. Chem. Soc.* **95**, 5033.

Yu, N.-T., and Liu, C. S. (1972). *J. Am. Chem. Soc.* **94**, 5127.

Yu, N.-T., Liu, C. S., and O'Shea, D. C. (1972). *J. Mol. Biol.* **70**, 117.

Yu, N.-T., Jo, B. H., and O'Shea, D. C. (1973). *Arch. Biochem. Biophys.* **156**, 71.

Yu, N.-T., Jo, B. H., Chang, R. C. C., and Huber, J. D. (1974). *Arch. Biochem. Biophys.* **160,** 614.

Yu, N.-T., East, E. J., Chang, R. C. C., and Kuck, J. F. R. (1977). *Exp. Eye Res.* **24,** 321.

Yu, T. J., and Peticolas, W. L. (1974). *Pept., Polypeptides Proteins, Proc. Rehovot Symp., 2nd Rehovot, Israel.* p. 370.

Yu, T. J., Lippert, J.L., and Peticolas, W. L. (1973). *Biopolymers* **12,** 2161.

Zagalsky, P. F. (1976). *Pure Appl. Chem.* **47,** 103.

Index

A

A_{2g} modes, 109
Absorptivity, 21
"Accordion" modes, 217
N-Acetyl glycine ethyl dithioester
  ($CH_3C(=O)NHCH_2C(=S)SC_2H_5$),
  structure of crystalline, 183
Acid side-chains, *see* Amino acids
Acoustical phonons, *see* Longitudinal
  acoustical modes
Actinomycin D, 206
Acyl chymotrypsins, 173–176
  information from RR, 175, 176
  structures investigated by RR, 173
Acyl enzymes, *see* Acyl chymotrypsins;
  Acyl papains
Acyl glyceraldehyde-3-phosphate dehy-
  drogenases, 176
  mixed population, 177
Acyl papains, 169–173, 177–183
  absorption spectra, 170, 178
  reaction scheme, 170, 178
  RR spectra, 171, 179
  structure, 170, 178, 182, 183
Adenine, 185, *see also* Bases
  relationship between absorption and RR
    spectra, 203, 204
  RR spectrum, 204

Adenosine triphosphate (ATP), metal binding, 188
Adrenodoxin, 151
Adriamycin, 205
α-helix, 73
  amide I, 77, 80
  amide III, 77
Amide, 73
Amide I mode, 32, 33, 77
  characterized polypeptides, 79
  nonprotein structures, 86
  protein, 80, 81
  resonance enhancement, 97
Amide I', 77, 85
Amide II, resonance enhancement, 97
Amide III mode, 77
  characterized polypeptides, 79
  protein, 80, 81, 86
  resonance enhancement, 98
  viral protein, 199
Amide III', 77, 85, 86
4-Amidino-4'-dimethylamino azobenzene,
  163
Amino acids, 72, 76
  acid side-chains, 90
  interactions in muscle proteins, 96
  interactions in virus, 200
Amphotericin B, 232
Amplitudes of oscillation, 15

Amylose, 233
Anharmonicity, 35
Anomalous polarization, 42, 106, 109
Antibody, 88
    antibody–hapten complexes, 163–166
    heterogeneity, 165
Anti-Stokes Raman scattering, 19, 23
Arsanilazocarboxypeptidase, 166–169
    absorption spectra, 168
    chromophore structure, 167
    RR spectra, 168
Ascorbate oxidase, *see* "Blue" copper
    proteins
Astaxanthin
    absorption spectrum, 127
    aggregates, 128
    excitation profile, 127
    RR spectrum, 128
    structure, 126
Azotyrosine, *see* Arsanilazocarboxypepti-
    dase
Azurin, *see* "Blue" copper proteins

B

Background luminescence, 62
Bacteriorhodopsin
    photolytic cycle, 139
    RR spectra, 139, 140
Base-pairing, 185–187
    infrared analysis, 32
    Raman detection, 189–191
    in viral RNA, 199
Bases
    isotope exchange, 188
    isotopic substitution, 188
    normal mode analysis, 188
    resonance Raman, 201
    quantitation by Raman, 188
Base stacking, 189–191
    in viral RNA, 199
Bence–Jones protein, 88, 90
Bending mode, 28, 30
β-pleated sheet, 73
    amide I, 77, 84
    amide III, 77, 84
    quantitation, 85
β-turn, 74
    amide I, III, 80, 86

"Blue" copper proteins
    charge transfer band, 151
    coordination site, 151
    function, 151
    RR spectra, 152
Boltzmann distribution, vibrational states,
    24
Bond polarizability theory, 43
Born–Oppenheimer approximation, 44
Boundary lipids, 228, 229
Brewster windows, *see* Plasma tube
Brominated bases, 205

C

Calcite wedge, *see* Polarization scrambler
Calibration lines, 56
Carbohydrates, 232
Carbon dioxide, normal modes, 28
Carbonic anhydrase
    function, 155
    inhibition by sulfonamides, 155–160
    tryptophan mode, 90
Carbon tetrachloride
    resolution test, 56
    spectrum, 41
    standard for depolarization ratios, 54
2-Carboxy-2'-hydroxy-5'-sulfoformazyl
    benzene, 163
β-Carotene, 36
    RR spectra, 15,15'-cis and all-trans, 132
    structure, 126
    triplet excited state spectrum, 67
Carotenoids, 125–133
    membrane probe, 230–232
    $\nu_{C=C}$, 127
    RR spectra *in situ*, 126, 129, 133
Catecholate RR peaks
    in enterobactin, 150, *see also* Dioxy-
    genases
Cell, liquid sample, 57
    capillary, 57
    flow, 60
    rotating, 58
    stirred, 58
Cellulose
    I and II, Raman spectra, 234
    polymorphic forms, 233

Center-of-mass coordinate, 16
Ceruloplasmin, *see* "Blue" copper proteins
Charge-polarization model
  carotenoid RR spectra, 130
  enzyme–substrate complexes, 169–176
  rhodopsin spectra, 136
Charge-transfer transitions
  ligand–metal, 99, 143–153
  between organic molecules, 142
Chlorophyll
  antenna, 122
  bacteriochlorophyll, 123–125
  cation, 125
  chlorophyll *b* structure, 123
  RR spectra, 122–125
Cholesterol, 223, 225, 226, 232
Chromatin, *see* Histones
Chromophore, 6, 9, 24
C–H stretching, phospholipids, 220–223
  I_{2885}/I_{2850}, 222
Chymotrypsin, 82, *see also* Acyl chymotrypsins
  α-chymotrypsin, 87
Cinnamic acid derivatives, *see* Acyl chymotrypsins; Acyl papains
C=N stretch, *see* Schiff base
Cobalamin, *see* Vitamin B_{12}
Coherent anti-Stokes Raman spectroscopy (CARS), 63
  resonance CARS, 63
Collecting lens, 53
Combination bands, 35
Corrin ring, 120
p-cresol, 38
C–S stretching, 88
Cytochrome *c*, 102
  absorption, 104
  excitation profile, 107
  excited state lifetime, 109
  ferrocytochrome *c* Raman optical activity, 70
  interaction with phospholipids, 227
  polarization dispersion, 110
  RR, 106, 112
Cytochrome oxidase
  interaction with phospholipids, 227
  photoreduction, 117
Cytosine, 185, *see also* Bases

D

Degenerate vibrations, 29
Degrees of freedom of motion, translational, rotational, vibrational, 27
(Deoxy)ribose–phosphate backbone, conformation, 193, 194
  MS2 virus, 199
  tobacco yellow mosaic virus, 200
5-Br-Deoxyuridine, RR spectrum, 206
Depolarization ratio ρ, 40
  measurement, 54
Deuterated lipids, 223–225
  selective deuteration, 224
  used to obtain DPPC–glycophorin temperature profile, 228
Diatomic molecule, vibrations, 11
  vibrational energy, 22
Dihydrofolate reductase, 161, 162
β-Dihydronicotinamide adenine dinucleotide (NADH), 87
4-Dimethylaminocinnamotyl imidazole, X-ray crystallographic analysis, 172
4-Dimethylamino,3-nitrocinnamic acid, X-ray crystallographic analysis, 173
4-Dimethylamino-3-nitrocinnamoyl papain
  RR spectrum, 171
  structure, 170
2,4-Dinitrophenyl (DNP) haptens, 164–166
  RR spectra
    bound to MOPC proteins, 165
    bound to rabbit antibodies, 164
Dioxygenases
  charge-transfer band, 148
  enzyme–substrate and enzyme–substrate–O_2 complexes, 150
  function, 148
  RR phenolate modes, 149, 150
Dipalmitoyl lecithin, *see* Dipalmitoyl phosphatidyl choline
Dipalmitoyl phosphatidyl choline
  C–H stretching region, 221
  effect of benzene, cations, cholesterol, and KI/I_2, 226
  experimental and theoretical melting curves, 220
  Raman spectrum as function of temperature, 218, 222, 223
  selective deuteration, 224
  sonication, 222

Dipalmitoyl phosphatidyl ethanolamine, 217
Disamycin A, 206
Displacement coordinate, 12
Dissociation energy, 35
Disulfide (–S–S–), 88, 94
Disulfide, preresonance, 97
Dithioacyl papains
  absorption maximum, 177, 178
  reaction scheme, 178
  RR spectra, 178, 179
  structure, 181–183
Dithioesters, *see also* Methyl dithioace-
    tate; Ethyl dithiohippurate
  delineation of isomer structure by X-ray
      analysis, 181–183
  models for enzyme-bound substrates,
      181
  use in monitoring catalysis, 177–183
DNA (deoxyribonucleic acid), *see also*
    Polynucleotide
  A form, Raman spectrum, 194
  backbone conformation, 193, 194
  B form, Raman spectrum, 194
  chemical constitution, 184, 185
  chemical modification, 197
  double helix, 187
  heavy-metal binding, 197
  histone binding, 200
  melting, 191
    Raman spectra, 192
DNP, *see* 2,4-Dinitrophenyl hapten
Drug–DNA complexes, 205–207
Drug–enzyme complexes, 155–162

E

Egg lecithin, 222, 224
Electric dipole transition moments, 43
Electric field, 17
Electric polarizability α, 17
Electronic transition moment, 44
Enzyme–substrate complexes, 169–183
Erythrocytes, *see* Membranes
Ethanolamine, 210
Ethyl dithiohippurate
  isomers, 180
  RR spectra
    as function of solvent, 180, 182
    as function of temperature, 180

E vector, 17, 39, 52
Excitation profile, 45
  adenine, 204
  ferrocytochrome c, 107
  measurement, 56
  oxyhemoglobin, 108
  providing excited-state lifetime, 109
Excited electronic state, 10, 22, 44
Excited-state lifetime *see* Excitation profile
Exciton coupling
  carotenoids, 128
  pheophytins, 125
Extrinsic membrane proteins, 211, 212, 227
Eye, lens protein, 89, 92

F

FAD, *see* Flavin adenine dinucleotide
Fatty acids, 209, 217
  deuterated, 223
    melting curve, 224
Fermi doublet, 38, 89
Fermi resonance, 35
Flavin adenine dinucleotide (FAD)
  bound to general fatty acyl-CoA dehy-
      drogenase, 142
  CARS spectrum, 65
  Raman, RR spectra, 141
Flavin mononucleotide (FMN)
  absorption, fluorescence spectra, 141
  Raman, RR spectra, 141
Fluorescence, 26
  absolute frequency fluorescence
      photons, 26
  time scale, 26
Fluorescence discrimination, 64
  time resolution, 64
  wavelength modulation, 66
FMN, *see* Flavin mononucleotide
Focusing lens, 53
Folate, 161, 162
Force constant, 12
Fossils, 70
Fourier transform infrared spectroscopy,
    lipids, 212
Franck–Condon factors, 45
Frequency, 4
Frequency doubling, 51, 202
Fundamental transitions, 36

Furylacryloyl derivatives, *see* Acyl chymotrypsins; Acyl glyceraldehyde-3-phosphate dehydrogenases

G

Gauche isomers, *see* Hydrocarbon chains
Gel phase, 211
Globular proteins, 74
Glucose, 233
Glyceraldehyde-3-phosphate dehydrogenase, 75, 176
Glycerol, 209
Ground electronic state, 10, 22, 44
Group frequencies, 29
Guanine, 185, *see also* Bases

H

Hammett constants, relation to RR spectrum, 160
Hapten, *see* Antibody
Harmonic oscillator, 13
Head-group vibrations, 213, 219
Heme absorption bands, 104
    Soret, α, β, charge-transfer, 105
Heme chemistry, 101
Heme photolysis, 115–117
    HbCO, 116
    Hb, HbO$_2$, 117
Heme proteins, 100, *see also* Cytochrome *c*; Hemoglobin; Myoglobin
    oxidation and spin states, 102
Heme RR, 104
    α–β excitation, 106
    anomalous polarization, 109
    iron–ligand modes, 117–121
    marker bands, 110, 114
    Soret excitation, 105
Hemerythrin
    function, 144
    $^{18}O_2$ and $^{18}O-^{16}O$ substitution, 146
    RR spectra, 146
Hemocyanin
    charge-transfer band, 144
    function, 144
    $^{18}O_2$ substitution, 144
    $^{18}O-^{16}O$ substitution, 146
    RR spectra, 144–146

Hemoglobin, 101
    excitation profile, 108
    Fe–O$_2$ stretch, 118
    oxygen coordination, 118
    RR, 106, 112
    R–T transition, 115
High-spin iron, 103
Histidine, 72, *see also* Imidazole
Histones, 200
    histone–DNA, Raman, 200, 201
    model for DNA complexes, 187, 201
    non-histone proteins, 200
    Raman spectra, 84
Hooke's law, 11
Hormones, 86, 88, 91
Hydrocarbon chains
    containing an olefinic linkage, 225
    gauche isomers, 210, 217–220
    interchain interactions, 221, 222
    order parameter, 219
    trans isomers, 210, 217–220
    vibrational spectra, 213–223
Hydrogen bonding
    in protein structure, 73
    between purines and pyrimidines, 31
Hydrophobic bonding, 156, 157

I

Imidazole, 90
Independent oscillator, 31
Induced dipole moment, 17, 21
Infrared intensities, 21
    determined by permanent dipole, 21
Infrared spectroscopy, 6
    comparison with Raman, 19
Inosine, 191
Inositol, 210
Internal coordinate, 16, 18
Intrinsic membrane proteins, 211, 212, 227
Inverse polarization, 42, 101, 109
Iron–sulfur proteins, 150, 151
Isoalloxazine, *see also* Flavin adenine dinucleotide; Flavin mononucleotide
    mode assignments, 142
    structure, 141
Isoenzymes, 91
Isotope shifts, 34
Isotopic substitutions, *see also* Deuterated lipids; Stearic acid-d_{35}

Isotopic substitutions (*cont.*)
  amide, D, 77, 85
  bacteriorhodopsin, D, 140
  bases, 188
  chlorophyll, ^{15}N, ^{26}Mg, 122
  2,4-dinitrophenyl haptens, D, ^{15}N, 165, 166
  hemerythrin, $^{18}O_2$, $^{18}O-^{16}O$, 144, 146
  hemocyanin, $^{18}O_2$, $^{18}O-^{16}O$, 144, 146
  isoalloxazine, 142
  methyl dithioacetate, D, ^{13}C, 180
  MS2 virus, D, 198
  myoglobin, $H_2^{18}O$, $^{18}OH^-$, $^{15}N_3^-$, [$^{15}N_2$]-imidazole, 119
  oxyhemoglobin, $^{18}O_2$, 118
  oxytyrosinase, $^{18}O_2$, 144
  poly(C)·poly(G), ^{18}O, 191
  protocatechuate 3,4-dioxygenase, $^{18}O_2$, $^{18}O-^{16}O$, 150
  retinal, D, 138
  stearic acid, 18-d_3, 221

K

Kinetic energy, 14

L

Laccase, *see* "Blue" copper proteins
Lagrange's equation, 14
Laser, 48
  argon lines, 50
  dye, 51
  krypton lines, 50
  plasma lines, 53
  pulsed, 51
Lateral crystal-like order, 222
Lecithin, *see* Phosphatadiyl choline
Ligand field energy, 102
Light, energy, 4, 51
Light wave, 17
Lipid bilayer, 210, 211
Lipid–protein interactions, 227–229, *see also* Cytochrome *c*; Cytochrome oxidase; Valinomycin
  glucagon–phospholipid, 227
  glycophorin–phospholipid, temperature profile, 228
  Gramicidin A–phospholipid, 227

insulin–sodium monodecylphosphate, 226
  lysozyme–phospholipid, 227
  lysozyme–sodium dodecyl phosphate
  melittin–phospholipid, temperature profile, 229
  myelin proteolipid apoprotein–phospholipid, 227
Lipids, *see* Phospholipids
Liposomes, 210
Liquid crystalline phase, 211
Liver alcohol dehydrogenase, 163
Lobster pigmentation
  blue protein, 129, 130
  infrared excited RR spectra, 131
  photon collection, 129, 130
  yellow protein, 129
Longitudinal acoustical modes (LAM), 214–216
  polyethylene, 216
Low-spin iron, 103
Lysozyme, 76, 81, 84, 85, 87, 226
Lysozyme, preresonance, 96

M

Melting behavior
  dipalmitoyl phosphatidyl choline, 218, 220
  dipalmitoyl phosphatidyl ethanolamine, 217
  using a deuterated probe, 223–225
  various phospholipids, 219
Membrane hysteresis, 230
Membrane potential, 231
Membranes
  erythrocyte, 230
  function, 208
  photoreceptor, 230
  rabbit thymocytes, 230
  RR probes, 230–232
  sarcoplasmic reticulum, 230
  structure, 208, 211, 212
Metalloporphyrins, 104, 105
Metalloproteins, 99
Methemoglobin, 102
Methotrexate
  bound to dihydrofolate reductase, 162
  RR spectra, 162
  structure, 161

Methyl dithioacetate, $CH_3CS_2CH_3$
  absorption spectrum, 25
  infrared spectrum, 20
  normal mode analysis, 180
  Raman spectrum, 20, 25
Methyl orange, 155
5-Methylthienylacryloyl chymotrypsin
  absorption maximum, 174
  RR spectra
    denatured and native, 174
    in D_2O, 176
Molecular energy levels, 21
  electronic transitions, 22
  vibrational transitions, 22
Monochromator, 48, 54
MOPC 315 and MOPC 460 immunoglobu-
  lins, 165–166
  DNP binding sites, 166
Morse equation, 35
Multichannel detector, 55, 66, 68, 94
Multichannel spectrometer, *see* Time-
  resolved Raman spectroscopy
Muscle proteins, 90, 94
Myoglobin, 101, 117, 119
  iron–ligand modes, 119–121

N

Nactins, 87
Neoprontosil, 159, 160
Netropsin, 206
Nicotinamide adenine dinucleotide
  (NAD^+), 86
Nitroanilines, absorption spectro, resonance Raman spectra, 6, 7
—N=N— bonds, twisting about, 163–165
—NO_2 group, RR features, 164–166
Non-histone proteins, *see* Histones
Non-totally-symmetric modes, depolarization ratios, 41
Normal coordinates, 27, 29
Normal modes of vibration, 27
Normal vibrations, *see* Normal modes of
  vibration
$\nu_{C=C}$, cis–trans isomers, 225
$\nu_{C=C}$ vs. $1/\lambda_{max}$ correlation, 128–129
  acyl chymotrypsins, 175
  acyl glyceraldehyde-3-phosphate
    dehydrogenases, 177
  rhodopsins, 136

ν_s^4 law, 24, 42
Nucleic acids
  absorption spectra, 203
  primary structure, 184
  resonance Raman, 201
  secondary structure, 184, 185
  tertiary structure, 185, 187
Nucleoside, 186
Nucleotide, 186
  metal binding, 188

O

Old yellow enzyme
  charge-transfer bands, 142, 143
  RR spectra phenol complexes, 142, 143
Olefinic linkage, *see* Hydrocarbon chains
—O—P—O— symmetrical stretch, *see*
  (Deoxy)ribose–phosphate backbone
Optical phonons, *see also* Transverse
  optical phonons
  polyethylene, 216
  skeletal, 218
Optic axis, monochromator, 49, 53
Oscillating dipole, 19, 53
Ovalbumin, 91
Overtone bands, 35
Oxidation-state markers, 111

P

Papain, *see* Acyl papains
Paraffins, 213, *see also* Hydrocarbon
  chains
Phase transition, 211
Phenolate ring vibrations, 149
Pheophytin, 122–125
  bacteriopheophytin, 123–125
  spectra, 124
Phosphate peaks, *see* (Deoxy)ribose–phos-
  phate backbone
Phosphatidyl choline, 209
Phosphoglycerides, *see* Phospholipids
Phospholipids
  C–H stretching region, effect of
    cholesterol and cations, 225
  LAM modes, 216
  1100–1150-cm^{-1} region, 217
  schematic, 210
  structures, 209, 210

Photoalteration parameter, 136
Photomultiplier, 55
Photon flux, 51
Photons, 1, 21
Photosynthesis, 122
π → π* transitions, 99
  carotenoid, 126–131
  corrin, 119
  heme, 104
pKs
  amino acid side chains, 72
  coupled, 160
pKs, macroscopic, microscopic, *see* pKs, coupled
Plasma tube, 52
Plastocyanin, *see* "Blue" copper proteins
PO_2^- symmetrical stretch, *see also* (Deoxy)ribose–phosphate backbone lipid, 218
Polarizability, *see* Electric polarizability
Polarization analyzer, 39, 54
Polarization of Raman scattering, 39
Polarization scrambler, 54
Polarized light, 39
Polarized π-electron system, 171, 172
Polarizer, 53
Polyatomic molecule, vibrations, 26
poly(dA) · poly(dT), 191
Polyethylene, *see also* Hydrocarbon chains
  phonon dispersion, 216
Poly-L-lysine, 79, 85
Polymethylenes, *see* Hydrocarbon chains
Polynucleotide, 187
  Raman spectra, 187–194
Polypeptide, random conformation, 78, 80, 85
Polypeptide, resonance Raman, 96
Polypeptide, skelatal mode, 78, 79
Polypeptide vibrations, 76, 79
Polypeptides, 71
poly(rA) · poly(rU)
  melting, 189
  Raman spectrum, 189
  RR spectrum, 205
poly(rA–rU) · poly(rA–rU), 190
  relationship between absorption and Raman spectrum, 203, 204
poly(rC) · poly(rG), 190
Potential energy, 13, 14

Potential energy function, 12, 13
Preresonance Raman spectra, 24
  frequency dependence, A term, 45
  frequency dependence, B term, 46
Protein
  conformation, 80
  denaturation, 75
  disordered structure, 84
  dynamics, 87, 90
  eye lens, *see* Eye
  low-frequency vibrations, 87
Protein structure, 71
  primary, 73
  quaternary, 75
  secondary, 73
  quantitation, 84–86
  tertiary, 74
Protocatechuate 3,4-dioxygenase, *see* Dioxygenases
Protoporphyrin IX, 100
Pteridine, *see* Methotrexate; Folate
Purine, *see* Adenine; Guanine
Purple membrane pigment, *see* Bacteriorhodopsin
Pyrimidine, *see* Cytosine; Thymine; Uracil
Pyrocatechase, *see* Dioxygenases

Q

Quantized energy states, 21–24
Quinaldine red, RR spectra bound to cells, 231, 232

R

Raman circular intensity differential (CID), *see* Raman optical activity
Raman difference spectroscopy, 91, 115, 222
Raman experiment, basic optics, 48
Raman hyperchromism, 190
Raman hypochromism, 190
Raman intensities, 42
  classical, 42
  quantum mechanical, 43
  A term, 44, 105
  B term, 44, 106
  simple rules, 46

Raman microprobe, 60
Raman microscope, 68
Raman optical activity, 70
Raman scattering, 1
  classical model, 17
  quantum mechanical model, 21
  time scale, 9, 23
Raman shift, 3, 5
Raman spectroscopy, 1
  advantages for biochemical studies, 8
  disadvantages for biochemical studies, 9
Raman spectrum, 3
  peak positions, 23
Rapid-flow technique, 60, 121, 135
Rapid Raman, *see* Time-resolved Raman
  spectroscopy
Rayleigh scattering, 19, 23
Reaction centers, 122–125, 132
Reduced mass, 15
Resonance Raman
  natural protein-bound chromophores, 99
  nucleic acids, 201
Resonance Raman labels
  membranes, 229–232
  nucleic acids, 205, 206
  protein sites, 154–183
Resonance Raman spectroscopy, 6, 24, 25
  advantages for biochemical studies, 9
Resonance Raman–X-ray approach to sub-
  strate conformation, 172, 173, 181–183
Retinal
  hydrogen out-of-plane bends, 138
  isomerisation, 134, 137
  RR spectrum *in situ*, 135
  structure, 133
Rhodopsin
  controlling photolability, 134
  dual-beam pump–probe method, 135,
    138
  light-induced cycle, 133, 134
  RR spectra
    bathorhodopsin, 136–138
    isorhodopsin, 136–138
    metarhodopsin I, 136–138
    metarhodopsin II, 136–138
    rhodopsin, 136–138
Riboflavin, 140
  RR spectra when bound to egg-white fla-
    voprotein, 142

Ribonuclease A, 84, 85, 89
  thermal denaturation, 91
RNA (ribonucleic acid), *see also* 5 S RNA;
  Transfer RNA; Polynucleotide
  backbone conformation, 195
  chemical constitution, 185
5 S RNA, 196
Rubredoxin, 150
Rule of mutual exclusion, 20

S

Sample problems, 61
Sampling techniques, 57
  temperature control, 61
Schiff base, 133, 138
  bacteriorhodopsin C=N stretch, 139,
    140
  rhodopsin C=N stretch, 136, 138
Serine, 72, 210
Simple harmonic motion, 15
Slit width
  mechanical, 55
  spectral, 55
Solids, sampling techniques, 60
Sonication, 222
Spectral sensitivity, spectrometer's, 56
Spectrometer, 54
  multichannel, 55, 67
  scanning, 55
Spheroidene in reaction centers, 132
Spike filter, 53
Spin-state markers, 112
Stearic acid-d_{35}, 223, 224
Stellacyanin, *see* "Blue" copper proteins
Stokes Raman Scattering, 19, 23
Stretching mode, symmetrical, antisym-
  metrical, 28, 30
Substrate–enzyme complexes, *see* En-
  zyme–substrate complexes
Sulfhydryl (–SH), 88, 94, 199, 200
Sulfonamides, 155–160
  azo-based sulfonamides
    absorption spectra, 156, 158
    RR assignments, 157
    RR spectra, 156–158
    structures, 156, 160
Symmetric modes, depolarization ratios,
  41

T

Temperature jump, 68
Tensor, 17
  antisymmetric, 109
Tensor invariants, 40, 42
  isotropic, 40, 53
  symmetric anisotropic and antisymmetric anisotropic, 40
Tetrapyrrole skeleton, 113
  core expansion, 112
  doming, 112–114
Thin films, interfaces, 60
Thiol *see* Sulfhydryl
4-Thiouridine, RR spectrum, 205
Thymine, 185, *see also* Bases
Time constant, setting, 56
Time-resolved Raman spectroscopy, 66
  heme photolysis, 116
Toxins, 86, 88, 89, 91
Transferrins
  charge-transfer band, 146
  Fe(III) sites, 146
  function, 146
  metal substitutions, 147, 148
  RR phenolate modes, 149
  RR spectra, 146–148
Transfer RNA
  aminoacylation, 196
  Raman spectra, 195, 196
  RR spectra, 205, 207
  secondary structure, 186
Trans isomers, *see* Hydrocarbon chains
Transition polarizability tensor, 43
Transition-state analog, sulfonamide, 157, 159
Transverse optical phonons, 214
Trypsin, 163
Tryptophan, 72, 90, 200
  H–D exchange, 90
  preresonance, 97
  resonance, 97
Tyrosinase
  function, 144
  $^{18}O_2$ substitution, 144
  RR spectra, 144, 145

Tyrosine, 72, 200
  doublet, 38, 89

U

Uracil, 185, *see also* Bases
  comparison Raman and RR spectra, 202
  hypochromicity, 189, 203
  relationship between absorption and RR spectra, 203
Uridine Raman, RR spectra, *see* Uracil
Uteroferrin, 148
  RR phenolate modes, 149

V

Valinomycin, 87
  interaction with phospholipids, 227
Vesicles, 210, 222
Vibrational coupling, 33
Vibronic transitions
  carotenoid, 126, 127
  heme protein, 105, 108
Viruses
  cucumber, 200
  fd, 200
  phage R17, 200
  Pf1, 200
  Raman spectra MS2 virus, 198, 199
  tobacco mosaic virus, 200
  turnip yellow mosaic virus, 199, 200
Vitamin B_{12}, 119

W

Water, 3
Wavelength, 2, 4
Wavenumber, 4, 56

Z

Zincon, *see* 2-Carboxy-2'-hydroxy-5'-sulfoformazyl benzene

Molecular Biology

An International Series of Monographs and Textbooks

Editors

BERNARD HORECKER

Roche Institute of Molecular Biology
Nutley, New Jersey

NATHAN O. KAPLAN

Department of Chemistry
University of California
At San Diego
La Jolla, California

JULIUS MARMUR

Department of Biochemistry
Albert Einstein College of Medicine
Yeshiva University
Bronx, New York

HAROLD A. SCHERAGA

Department of Chemistry
Cornell University
Ithaca, New York

HAROLD A. SCHERAGA. Protein Structure. 1961

STUART A. RICE AND MITSURU NAGASAWA. Polyelectrolyte Solutions: A Theoretical Introduction, *with a contribution by Herbert Morawetz*. 1961

SIDNEY UDENFRIEND. Fluorescence Assay in Biology and Medicine. Volume I—1962. Volume II—1969

J. HERBERT TAYLOR (Editor). Molecular Genetics. Part I—1963. Part II—1967. Part III—Chromosome Structure—1979

ARTHUR VEIS. The Macromolecular Chemistry of Gelatin. 1964

M. JOLY. A Physico-chemical Approach to the Denaturation of Proteins. 1965

SYDNEY J. LEACH (Editor). Physical Principles and Techniques of Protein Chemistry. Part A—1969. Part B—1970. Part C—1973

KENDRIC C. SMITH AND PHILIP C. HANAWALT. Molecular Photobiology: Inactivation and Recovery. 1969

RONALD BENTLEY. Molecular Asymmetry in Biology. Volume I—1969. Volume II—1970

JACINTO STEINHARDT AND JACQUELINE A. REYNOLDS. Multiple Equilibria in Protein. 1969

DOUGLAS POLAND AND HAROLD A. SCHERAGA. Theory of Helix-Coil Transitions in Biopolymers. 1970

JOHN R. CANN. Interacting Macromolecules: The Theory and Practice of Their Electrophoresis, Ultracentrifugation, and Chromatography. 1970

WALTER W. WAINIO. The Mammalian Mitochondrial Respiratory Chain. 1970

LAWRENCE I. ROTHFIELD (Editor). Structure and Function of Biological Membranes. 1971

ALAN G. WALTON AND JOHN BLACKWELL. Biopolymers. 1973

WALTER LOVENBERG (Editor). Iron-Sulfur Proteins. Volume I, Biological Properties—1973. Volume II, Molecular Properties—1973. Volume III, Structure and Metabolic Mechanisms—1977

A. J. HOPFINGER. Conformational Properties of Macromolecules. 1973

R. D. B. FRASER AND T. P. MACRAE. Conformation in Fibrous Proteins. 1973

OSAMU HAYAISHI (Editor). Molecular Mechanisms of Oxygen Activation. 1974

FUMIO OOSAWA AND SHO ASAKURA. Thermodynamics of the Polymerization of Protein. 1975

LAWRENCE J. BERLINER (Editor). Spin Labeling: Theory and Applications. Volume I, 1976. Volume II, 1978

T. BLUNDELL AND L. JOHNSON. Protein Crystallography. 1976

HERBERT WEISSBACH AND SIDNEY PESTKA (Editors). Molecular Mechanisms of Protein Biosynthesis. 1977

TERRANCE LEIGHTON AND WILLIAM F. LOOMIS, JR. (Editors). The Molecular Genetics of Development: An Introduction to Recent Research on Experimental Systems. 1980

ROBERT B. FREEDMAN AND HILARY C. HAWKINS (Editors). The Enzymology of Post-Translational Modification of Proteins, Volume 1. 1980

WAI YIU CHEUNG (Editor). Calcium and Cell Function, Volume I: Calmodulin. 1980. Volume II. 1982

OLEG JARDETZKY and G. C. K. ROBERTS. NMR in Molecular Biology. 1981

DAVID A. DUBNAU (Editor). The Molecular Biology of the Bacilli, Volume I: *Bacillus subtilis*. 1982

GORDON G. HAMMES. Enzyme Catalysis and Regulation. 1982

GUNTER KAHL and JOSEF S. SCHELL (Editors). Molecular Biology of Plant Tumors. 1982

P. R. CAREY. Biochemical Applications of Raman and Resonance Raman Spectroscopies. 1982

In preparation

CHARIS GHELIS and JEANNINE YON. Protein Folding. 1982

OSAMU HAYAISHI and KUNIHIRO UEDA (Editors). ADP-Ribosylation Reactions. 1982

WAI YIU CHEUNG (Editor). Calcium and Cell Function, Volume III. 1982

ALFRED STRACHER (Editor). Muscle and Non-Muscle Motility, Volume 1. 1982